NO SAFE SPACES

A YOUNG MARINE IN VIETNAM

ANTHONY H JOHNSON

DRAGON PUBLISHING

This work is dedicated to the bravest and toughest men I have ever known. The Marines who served in Vietnam, and most especially, those who died there. Although these stories are about my unit, the 1st Battalion of the 7th Marine Regiment, all Marine infantry units have similar stories of bravery and sacrifice.

Members of the other services are recognized for their service as well. It takes a team effort to fight a war.

This book is also dedicated to the late Mrs. Catherine King, my high school English teacher.

INTRODUCTION

These stories are not just about war and history, they are about life. The participants of the actions described here were very young. Many had dropped out of high school and sought the stability and structure of life in the Marine Corps or Navy. What they did should be recorded for history, and they should be honored for their courage and sacrifice.

I hope that these stories will be of value to teachers, coaches, scoutmasters, counselors, and others who deal with youth. The teacher who sees a young person struggling with schoolwork should understand that inadequate performance does not necessarily indicate an inability or an unwillingness to learn. They may be dealing with issues that are unknown to you and could benefit from your support.

Teachers should see the potential in students who are having difficulty with their schoolwork. The kid who is failing an eighth grade civics class might become a member of congress or judge one day if he or she can get past this rough spot in life.

You cannot use your difficult conditions as an excuse to not do the best you can. We don't all start from the same place, but we are all held to the same standard. That may not seem fair. Life is not fair, so you might as well get used to that now and adjust to that reality. You have to play the hand you are dealt, not the one you wish you had. Energy wasted on brooding over your situation, or complaining about it, could be better used in dealing with it. Don't let people around you discourage you. Keep moving forward. Opportunities will present themselves. You may have to work harder than others, but you can do it.

The Vietnam war was complicated. I will leave the political and social commentary of that war to others. I have made every effort to be straightforward and factually correct in what I have written. My memory of the event of 50 years ago has been supplemented by my reading of official records of the engagements that I describe.

I have tried my best to report these events as accurately as possible. My memory of some of these events is clear even fifty years later. Some of the details were reconstructed by my recent conversations with Marines who participated in the events described. Even so, it is important to note that if you ask two participants in the same firefight to describe it immediately afterward, you might get two different versions. Neither person is lying, they just saw things from different views. Sometimes, they were in different parts of the fight.

I hope that you will enjoy these stories and that they will bring you a better understanding of the what combat in

Vietnam was like and those who fought there. History should be adequately told, the young men should be honored for their courage and sacrifice, and young people should be inspired and encouraged.

Each of the stories in this collection was written to stand alone and several have been published individually. The reader will notice that some information is repeated in more than one chapter and may seem redundant. The repetition provides context to the individual story that would otherwise be lost. I apologize for the need for that.

BEFORE THE CORPS

In the spring of 1966, I was a 17-year-old high school junior. I would have been a senior if I had not had to repeat the 8th grade. I lived on my own and had for a couple of years. "Home" was the back of a funeral home, where I worked up to 120 hours a week. The 48 hours a week that I had off included the 40 hours that I was required to attend school. I could sleep during night-time hours if things were quiet, but I was subject to being called for ambulance duty or deaths in the middle of the night.

The benefits included room but no board. I bought my own meals, clothing and other necessities with the $19.20 I was paid each week for working as a funeral home assistant. I shared a room in the back of the funeral home with others who worked there. In those days, funeral homes provided ambulance service. That made sense because they had the only vehicles in small towns that were configured to transport someone lying down. A hearse could be configured to carry a gurney or a casket. Between funerals, the stretcher was fitted, and a removable rotating red light attached to the roof.

Minor injuries could be treated in one of the two small hospitals in my town of DeFuniak Springs, Florida. More severe injuries required the patient to be transported 82 miles to better-equipped hospitals in Pensacola. There were no interstate highways at the time. The ambulance trip to Pensacola would take several hours, in addition to the time spent transporting the patient from an accident scene to a local hospital, where the decision was made that the only acceptable level of care could be provided in Pensacola. Sometimes I made it back "home" in time for school. Often, I was late. If I made it to school, I was exhausted and not in a mental state conducive to learning.

I had no complaints. I was happy to have the job. I had previously worked as a paperboy for the Pensacola News Journal, which arrived every morning at about 3 a.m. at the bus station in DeFuniak Springs. I would roll or fold the papers, depending upon their size, and deliver them along my 6-mile route. I would complete the 12 miles out and back before most people woke up. Then, I would go to school. After school, I would go back for the afternoon paper. I did that every single day, 365 days a year, no matter what the weather. On weekends, I would visit each of my customers and collect their subscription fees.

I had also worked as a shoeshine boy in a barber shop, where I made 25 cents per shine. Later I worked for a grocery store where I stocked shelves, kept the store clean, and delivered groceries on my bicycle.

My easiest job was that of an apprentice bootlegger. My grandfather, who had retired from the Florida State Road Department as a laborer, supported his tiny pension by smuggling illegal moonshine whiskey. I was his assistant. We would drive to the rural areas of North Florida where he would pick up a load of moonshine and conceal it under a tarp in the back of his 1950 green Ford pickup truck. Once we got it home, it was my job to siphon the whiskey from 10-gallon jugs into quart mason jars, using a length of garden hose. I learned to drive in that truck, loaded with "white lightning," on the back roads of Walton County. Of course, I was too young to have a driver's license.

My first job was selling boiled peanuts. My grandfather grew the peanuts on his 20-acre farm on the outskirts of town. I helped my grandmother boil them and put them in small paper bags. I rode my bicycle around town selling bags of peanuts for 10 cents each. I was eight. This was a lesson to me in business. At the end of the summer, I had just enough income to buy a $20 US Savings Bond. It would have been inconceivable to my grandparents or myself that I would use that savings bond to help pay my law school tuition 20 years later.

As a kid, I had little opportunity for recreation or activities outside school and work. The only money I had was what I earned at my various jobs. I attended only one school prom. That was because the homecoming queen asked me to escort her. In other years, my meager budget would not support the purchase of a sport coat and tie necessary for those occasions. Later, I would buy a black suit for use at my funeral home job.

I spent almost all of my youth in the custody of my grandparents. My single mother was a school teacher in Pensacola and found it difficult to provide adequate childcare. When I was 15, my grandfather suddenly died. I knew that my grandmother would have difficulty providing for me on her social security income alone. It was then that I left my newspaper route and took the job at the funeral home.

The only clothing I bought, other than for work or school, was a Boy Scout uniform. I joined the Boy Scouts at the earliest age permissible. Scouting was the only activity I was able to participate in because of school commitments and the need to work. Like many other kids in the rural south, I enjoyed being outdoors. Scouting skills came easy to me, and I moved up through the ranks. One summer, I was able to get a week off work to attend Boy Scout camp. It was the best summer of my childhood. I earned merit badges in canoeing, rowing, swimming, and lifesaving. I also earned a badge for swimming a mile. I was an expert marksman with a rifle, having been taught to shoot by my grandfather, who had served as an Army infantryman in World War I and thought that every boy should know how to shoot. I eventually became an Eagle Scout. It was the first time I had ever excelled at anything. It was the first time that I had the opportunity to excel.

The Boy Scouts also taught me a valuable life lesson. If you want to accomplish something, set the goal, determine the steps necessary to accomplish it, make a plan to execute those steps in the proper sequence, get started and don't stop till you accomplish your goal. Stay focused and keep going. Becoming an Eagle Scout does not require you to be a super kid. It does require you to focus and be disciplined and to work. Someone who becomes an Eagle Scout has learned a valuable lesson in how to succeed, not only in that endeavor, but in several other areas of life.

The Boy Scouts also instill the values of honor, integrity, honesty, keeping your word, and treating others with kindness. Those lessons would benefit me throughout my life.

I was trying hard to recover from my academic setbacks. I attended summer school one year to make up some of my failed classes and took an extra class in my junior year. If I passed all my classes, I would be able to graduate with my original class. One of my teachers, Mr. Taylor, told me that I needed to "buckle down" in his class because I was so far behind that I would need to make a 95 on the final exam to pull my grade average up to passing. I told him I would bet him an RC Cola and a moon pie that I would pass. I studied hard for the exam and made a 92. Three points short. I knew that the teacher was strict and that I had flunked the course, which meant that I would not be graduating. It was not the only class that I was flunking. My English teacher, Mrs. King, had informed me that I was behind in her class and in danger of failing. In fact, I was flunking.

The television news stations, all three of them that could be received in DeFuniak Springs, kept us informed about a war building up in a place called Vietnam. The news stories informed us that the communists in North Vietnam wanted to take over the free country of South Vietnam. The commies were committing all kinds of atrocities against the freedom-loving but defenseless villagers of South Vietnam, taking the rice crops that the farmers depended upon for their livelihood, forcing the young men to labor and fight for them, and bullying the local population. Village chiefs who stood up to the communists, known as Viet Cong, were assassinated.

The news casters reminded us about the "domino theory." If the communists kept taking small countries, one at a time, eventually they would all fall and we would be next. "Better to fight them in Asia than wait till they come ashore in Pensacola" was the logic. It was only 20 years after World War II had ended and there was still a patriotic feeling in the country. Movie theaters featured films depicting the heroic actions of our military in World War II. The John Wayne flick, "The Sands of Iwo Jima," was a personal favorite. The power of the media was focused on preparing us for a war.

Young men were required to register for the draft. Military recruiters stalked high schools, looking to fulfill their quotas. The recruiters seemed to compete with each other over who could provide the most enticing pitch. "Join the Navy and see the world" had some appeal to teenagers who had never been 100 miles outside Walton County. The Air Force recruiters promised that you would learn valuable job skills if you jointed them. Hell, the Air Force would probably make you an astronaut, "just sign here." Uncle Sam pointed his finger at you from a poster with the words, "I want you for the U.S. Army." Uncle Sam did not have to say that the U.S. Army would take you, whether you wanted to join or not, if you were drafted.

One military branch took a different approach. The Marine Corps did not even bother to send a Marine. A Navy Captain spoke at our high school. He said that if you joined the Marines, they would not promise you anything but "a rifle, a pack, and a hard time." The Marine Corps was the branch of service that almost guaranteed that you would be trained for and participate in combat. No sightseeing trips or sophisticated jobs skills here. Just hardcore combat skills and the chance to use them. It was the best chance to step up and help those villagers in Vietnam who were being abused by the communist North Vietnamese, who wanted to conquer the free people of the south.

Reports from the war in Vietnam were appearing on television and newspapers. The Marine Corps stood apart from the other branches by seeming to inflict the most damage to the enemy while taking the most casualties. At the funeral home where I worked, we received the body of an American killed in Vietnam. He was a Marine who died of a gunshot wound. The cars of the recruiters could sometimes be seen in town. They all bore the emblems of their respective services. The Army, Navy, and Air Force emblems on those cars gave rise to the impression that those services offered a way out of my small town and the gateway to a better future. The impact of viewing those emblems, however, had no more impact on me than seeing the sign identifying a post office. The sight of the Marines' eagle, globe and anchor emblem, the same one worn by John Wayne at Iwo Jima, had a much more powerful impact. Just the sight of that iconic emblem quickened the pulse and gave rise to a sense of extreme challenge.

The boys who would be leaving Walton high school, one way or the other, at the end of the school year had decisions to make. At that time there were no local junior colleges. The colleges were all over 100 miles away, making it necessary to move to those college towns and live there while you attended college. Of course, you had to have the grades to get into one of them. The boys who were not able to attend college, which was most of us, knew that we would find ourselves in the military soon. The goal of many was to minimize the discomfort and danger associated with that service. They usually joined the Navy and Air Force, both of which had limited involvement in the ground war in Vietnam. If you were lucky, you could stay in the United States, perhaps at a base near enough to drive home to DeFuniak Springs on weekends.

Although I saw little point in attending the rehearsal for the graduation ceremony, I went. The graduating class was sitting in the bleachers at the football field when the principal read off a short list of students who would not be graduating with the class. To my great surprise, my name was not on it. I asked, "Sir, is my name on the list?" Looking as surprised as I was, he reviewed his list saying, "Didn't I call your name?" The principal, whose office I had often visited, actually looked disappointed as he turned to Mr. Taylor and asked, "did Tony pass your class?" Mr. Taylor responded, "He got a 69 and a half," as he passed up a small brown bag containing an RC Cola and a moon pie.

The graduating class of Walton High School stood and applauded. I was actually a high school graduate. In any case, I would soon be leaving DeFuniak Springs six weeks after my graduation ceremony.

I had joined the Marines.

HELLO VIETNAM

When the rear ramp of the C-130 opened, Vietnam rushed in. The heat and smell of the country were in sharp contrast to the cooler, cleaner air that had been trapped in the aircraft passenger compartment during our transit from Kadena Air Base in Okinawa. Marines, most of them young, sat on the canvas seats that lined both sides of the aircraft. The seats were not padded, and the jarring ride had not been kind to the sore buttocks of every Marine on the plane.

Each of us had spent two days in Okinawa being processed into the Fleet Marine Force, Western Pacific, or in Marine speak, "WestPac." That processing included a painful gamma globulin injection which, because of the size and viscosity of the dose, was administered to the buttock. The dreaded shot was supposed to boost our immune systems, but it meant that all Marines arrived in Vietnam with a sore posterior, especially those that came on C-130's.

We each had a sea bag containing most of what we owned. The rest of our belongings, including all our uniforms except for one set of khakis and a couple of cotton stateside utility uniforms which we had with us, were stored in a big warehouse in Okinawa. Stacked neatly several stories high, showing our names and serial numbers on the ends, they would be easy to identify and remove if we were not coming back for them.

It was around noon on an early November day when the human cargo of the C-130 filed out onto the tarmac in Da Nang with our sea bags. We were directed to a building with a tin roof and no walls, where a Marine with a clipboard led us to the 1st Marine Division processing on one side of the building or the 3rd Marine Division on the other side.

I carried my sea bag to the shade of the building on the 1st Marine Division side, along with several others. It was there that another Marine with a clipboard informed us of our various unit assignments. I was assigned to the 7th Marines. We were told that our regiments would be sending transportation for us and ordered to wait at the spot where we were to be picked up.

I waited, sitting on my sea bag in the shade of the building while vehicles arrived. I watched as several vehicles, primarily the heavy 6x6 trucks, stopped and announced the units that sent them. I watched as the new arrivals climbed on board the trucks and set off for their new assignments. The vehicles from the units assigned to the area of the headquarters of the 1st Marine Division in Da Nang were well cared for. Those from more outlying areas, like the infantry regiments, showed more wear. I had been waiting for some time when a truck approached that appeared to have led a hard life. It was filthy with red dust and showed visible battle damage.

Please God, don't let this be the truck from the 7th Marines! The truck came to a stop at the pickup point, and the driver yelled, "Who's here for the 7th Marines?" A couple of other new arrivals and I climbed into the bed of the truck where a Marine in a flak jacket and helmet waited. He handed us M16 rifles and asked if we knew how to use them. The M16 had recently replaced the M14 in the 7th Marines. He said, "You may need these."

As our truck rolled out of the airbase and into the city of Da Nang, we passed through "Dog Patch," a residential area of ramshackle buildings along the road leading out of town. Vietnamese people with conical straw hats and what seemed to be pajamas went about their business. They rode bicycles and mopeds, pulled rickshaws and walked. Soon, we were on an unpaved road of red dirt heading south out of the city. Farmers in rice fields carried huge loads on each end of a pole balanced on one shoulder as they shuffled along the roadside.

A common sight was the Bubalus Bubalis, the Vietnamese equivalent of a tractor. More commonly known as the Water Buffalo, this animal can weigh as much as 1,200 pounds and be over 6 feet tall at the shoulder. Children often rode or led these huge beasts as they pulled heavy farm implements through the rice fields.

The Marine who had come with the truck stood behind the cab looking forward and resting his M16 on the top of the cab as his eyes scanned the road ahead and the rice fields on either side. There were two primary dangers in traveling by truck along these roads: mines and ambush, often at the same time. Each morning, a mine sweep team cleared the roads before vehicles were allowed to travel, but sometimes, the sweep teams missed a mine. There were also plenty of opportunities for Viet Cong to plant a mine after the sweep team had cleared the road.

The mines could be activated by the pressure of a vehicle running over them, in which case the VC might be nowhere around when the device detonated. Mines could also be command-detonated by a VC who was watching the traffic from a hiding place and deciding when to trigger the explosion. Command-detonated mines allowed the VC to be more selective in their target selection and avoid accidentally blowing up a water buffalo that stepped in the wrong place in the road.

Ambushes were not uncommon. The VC knew that Americans could always be found using the roadways to transport material and personnel. Enemy troops, sometimes dressed as peasants, could open fire on a passing truck from concealed positions along the road, especially when trucks slowed or stopped. That is why Marine trucks did not slow or stop. The locals all knew to move to the side of the road when a truck approached. If they were in the roadway and moved slower than the truck, the truck driver would sound his horn in warning but would not slow down. More than one Vietnamese bicycle became a hood ornament on a Marine 6x6 truck. The truck drivers did not want to injure the Vietnamese but wanted less to risk ambush by making themselves and their trucks vulnerable.

The most deadly danger was a combination of command-detonated mine and ambush. Hidden wires from the mine would be connected to a handheld control switch, which would send an electrical current to the "blasting cap" of the mine. In that scenario, the enemy would explode a mine just as the target passed over it. An enemy force would then open fire on the damaged vehicle with automatic weapons and rocket-propelled grenades to kill the stunned and injured survivors.

Hill 55

After an hour of travel through the rice fields and occasional villages, the truck came to a large mound of red dirt surrounded by rolls of barbed wire known as concertina wire. Around the perimeter, inside the concertina wire, were bunkers made of sandbags, including the roofs. The fortification was for a good reason. Only three months earlier, the enemy had attacked and overrun the hill.

A Marine with a rifle stood by a guard shack at the entrance to the compound and motioned the truck forward as we approached the opened wooden gate, which was also covered with barbed wire. A wooden sign announced that we had arrived at Hill 55, home of the 7th Marines.

The Marine in the back of the truck collected the M16's he had loaned us as we offloaded in front of a building with plywood sides and a tin roof. Inside, we presented our orders to a clerk, reporting for duty with the regiment. We were sent to chow at the plywood mess hall and provided a bunk in an open squad bay, located in another plywood building with a tin roof.

During chow, we sat among other Marines who were assigned to Hill 55 and took the opportunity to gain information about the regiment, the units where we might be attached, and additional information that might be helpful, especially if we were to have choices to make.

The 7th Marines was one of the infantry regiments of the famed 1st Marine Division. The regiment had a distinguished history, having served well at places like Guadalcanal in World War II and the Chosin Reservoir in Korea. We were proud to become members of such a regiment and understood our duty to honor that history and continue its reputation. Maintaining an honorable tradition was important to Marines, having been instilled in us since boot camp.

We were all anxious to learn what our assignments would be. There were three battalions in the regiment, designated simply as the 1st, 2nd and 3rd. Each of the battalions had four companies identified by letter beginning with A or Alpha. In addition to the rifle companies, each battalion had a Headquarters and Support company, known as H&S.

After our first wakeup in Vietnam, we had breakfast and reported to the admin building as instructed. Since our arrival, the staff had been deciding where each of us, based upon the needs of the different battalions and our military occupational specialties (MOS), would be assigned. My assignment was the 1st Battalion.

1st Battalion Hill 10

The 1st Battalion, 7th Marines, expressed simply "1/7," was based on Hill 10, between Hill 55 and Da Nang. The mission of 1/7 was to protect Da Nang. The North Vietnamese Army (NVA) and Viet Cong (VC) utilized Russian-built 122mm rockets as their artillery. The range of the rockets allowed the VC to attack the vital complex in Da Nang, including the sprawling airbase, from over 12 miles out. The area from where those rockets were launched was known as the "rocket belt."

The units of 1/7 were distributed on small combat bases from which they patrolled the rocket belt. Delta Company was on Hill 41, another rifle company was on Hill 22, and two companies were housed along with battalion headquarters on Hill 10.

On my second day in Vietnam, a truck dropped me off at Hill 10, where I reported to battalion headquarters to be assigned to whatever unit within 1/7 fate would send me.

0351

My military occupational specialty code was 0351. I was trained as a jack of all trades in the infantry. In addition to being a basic infantryman, as were all Marines, I had been trained in demolitions, flamethrowers, 3.5-inch rockets, and the 106 mm recoilless rifle. Flamethrowers were not used much in Vietnam although some 0351's were still assigned to that weapon. The 3.5-inch rocket, which some would describe as a bazooka, had been rendered obsolete by the disposable, one-shot, light anti-tank weapon, or LAW. Most demolition tasks were handled by engineers. The 106 mm recoilless rifle, however, was very much in use. In fact, I could not help but notice one mounted on the top of a 70-foot observation tower on Hill 10.

The 106 mm Recoilless Rifle

The M40 106 millimeter recoilless rifle was developed in the early 1950's. Its primary purpose at the time was as an anti-tank weapon, but it could also be used in an antipersonnel role. It is a single-shot, breech-loaded, direct-fire weapon with a maximum effective range of 1,100 meters. A .50 caliber spotting rifle is coaxially mounted on the top of the barrel and bore sighted with the main barrel to ensure that the spotting round will impact precisely where the 106mm round will impact. The spotting round is a tracer with a white phosphorous tip, which produces a white puff of smoke upon impact.

At 462 pounds, its portability is limited. It can be delivered by helicopter or truck and quickly set up in the field or mounted on mobile platforms like tracked vehicles, trucks or jeeps. It is a crew-served weapon with a team of three: a gunner, assistant gunner and someone to direct fire.

The term "recoilless" refers to the fact that the rifle, which sits on a tripod, has a breech that is open to the rear and expels gas from the explosion of the powder charge when the gun is fired. It is the same principal as a bazooka; there is a back-blast but no recoil. When the weapon is fired, fire comes out the muzzle in the form of "muzzle blast," and a huge "back blast" comes out the back. The back-blast can be fatal to anyone standing behind the weapon. The first time I fired a 106 recoilless rifle, I thought it had exploded. Flame came out the muzzle and the breech, wrapping around me. It was so loud that it was physically painful. My ears felt a sharp pain. I have not heard anything in the high frequencies since that time in 1967.

It is a big rifle, which fires a large round with an explosive warhead. It is a "point and shoot" weapon, in contrast to artillery, which requires time to compute the numbers for locations radioed to them. The gunner aims and fires it directly at the target, getting a round off within seconds of identifying the target. The Marines cleverly re-purposed the anti-tank weapon to counter rocket artillery used by the NVA and VC.

A rocket attack lasts only a very few minutes. The rocketeers know that their position has been broadcast to every American within 20 miles by the brilliant trails of fire streaking almost a thousand feet into the night sky by the departing rockets.

Observers of the rocket launch could estimate the position of the launch site on a grid map and convey the information to an artillery battery. The artillery battery converts the coordinates to settings for their guns, which are then applied. By the time the first artillery round lands, the attack is over, and the enemy rocketeers have fled. The 106mm recoilless rifle gunner observing a rocket launch can get a round on the way to the launch site within seconds, while the launch site is active and providing an excellent visual target.

To maximize the effectiveness of the 106 mm rifles, Marines mounted them atop observation towers as high as 70 feet. That not only provided a 360-degree field of fire, but it also increased the range and gave the crew an excellent observation point.

The Marine Corps was known for its "flexibility" in assignments, which meant that it was not uncommon for a Marine to spend months being trained in a skill and find himself assigned to a completely different role. I had no idea what to expect as an assignment, other than that it would be in the infantry.

The basic Marine rifleman, the backbone of the Marine infantry, was assigned the military occupation specialty code (MOS) of 0311. After being trained as a basic rifleman, some were selected for additional specialized infantry training. Those qualified to operate the M60 machine gun were assigned the MOS 0331; mortar men were 0341. Any MOS starting in 03 was infantry.

Marines who had been trained in the special infantry weapons were often sent to rifle companies to fill manpower gaps in platoons that were understrength because of high casualty rates. Every unit in 1/7 was significantly under strength and badly in need of 0311 riflemen.

106 Platoon

Newbies go first to the battalion headquarters on Hill 10, where they are assigned to a company. Some stay on Hill 10 with the two companies assigned here. There are two other hills, 41 and 22, each with a company of Marines. The trucks from each of the outlying hills come to Hill 10 to pick up stuff, food for example, and take it back to their hills.

In addition to its four rifle companies, 1/7 had a Headquarters and Support (H&S) company. The various weapons platoons were assigned to H&S. The members of that platoon could then be attached to wherever the need for their specialty was greatest. Navy Corpsmen were also assigned to H&S to be allocated where they were most needed. I was assigned to the 106 mm recoilless rifle platoon in H&S Company. The 106 platoon would send me wherever a 106mm crew person was needed most.

I reported to the 106 mm recoilless rifle platoon located in a bunker made of sandbags and wooden ammo crates filled with sand. The senior Marine there was a staff sergeant. I was sent to supply to draw a rifle, jungle uniforms with boots, and "782 gear." Marines refer to their helmets, flak jackets, ammunition belts, canteens, packs and other similar gear as "782 gear" after the Department of Defense form 782 on which they sign for that equipment.

I stood in line with other newbies at the supply bunker on Hill 10. There was a dirt walkway leading to the bunker and weeds growing on the sides of the walkway. I noticed several articles lying in the weeds. One was a helmet with a camouflage cover. The bottom of the helmet faced up so that I could see into it. It had a hole the size of a bullet in one side and a pool of blood coagulated in the bottom. Flies buzzed around it. Nearby were a couple of bloodstained flak jackets, also attracting flies. The reality of war was becoming clear.

As I waited for word on my assignment, I asked the Marines around me about the various units of 1/7, where they were, and which would be the best and worst assignments. Hill 10 was the largest of the bases with two rifle companies, the headquarters element, and an artillery battery.

It was also the most secure. There were two remote bases, Hill 22 and Hill 41, each having one rifle company. Hill 41 could be seen from Hill 10 and was the patrol base for Delta Company. The consensus of opinion was that duty on Hill 10 was preferable, Hill 22 was the next best choice, and Hill 41 was to be avoided. "They get hit every night" was the comment one Marine made about Hill 41.

The only place more dangerous than Hill 41, by consensus of opinion, was a Combined Action Platoon (CAP) where Marines lived in the village with local Popular Forces (PF's) as advisors. There were six CAP units within 1/7's area of operations, one of which had a 106mm recoilless rifle assigned to it. Each CAP unit had only a squad of a dozen Marines and a platoon of PF's. They were often attacked and sometimes completely wiped out. Fortunately, new arrivals were not assigned to the CAP unit. That left the most dangerous place that I could be assigned as Hill 41 with Delta Company.

After morning chow at the mess hall on Hill 10, I returned to the 106 "hootch" (tent) where the Staff Sargent told me to catch the daily supply run truck to Hill 41. As fate would have it, I was assigned to Delta Company.

Every morning, a truck arrived from each of the two outlying companies of 1/7 to pick up supplies and transfer personnel. Tuesday was the day that new arrivals boarded the trucks on Hill 10 to be taken to their assignments on Hills 41 and 22. There were several of us newbies on the truck on that particular Tuesday. Delta Company, like the other Marine line companies, had a high attrition rate.

• • •

Hill 41

After about a 20-minute truck ride, we arrived at the gate of Hill 41, which would be my home for several months. The truck stopped just inside the entrance to discharge its passengers. I climbed down from the bed, pulling my sea bag with me, and looked around at my new neighborhood.

Like other bases in the area, the hill was a mound of red dirt surrounded by rows of concertina wire and ringed by bunkers made of sandbags. There was a 70-foot tower that had been determined to be too rickety for a 106 to be mounted on it. Therefore, the 106 was mounted on one of the sandbag perimeter bunkers. The place looked desolate and stark. There was almost no vegetation, thanks to the Dow Chemical Company product known as Agent Orange, which had removed any foliage that could hide an enemy who might make it past the concertina wire.

It was the base of Delta Company, known as the "Delta Death Dealers." Delta Company was known for leaving the ace of spades playing cards on the bodies of dead enemy soldiers. They even had special cards printed with their unit name. In addition to the three platoons of 0311 riflemen, there was the 106 mm recoilless rifle unit of which I would be a member, an 81mm mortar team, 0331 machine gunners and Navy corpsmen. (The Marine Corps does not have medical personnel, so those services are provided by the US Navy.) The corpsmen and machine gunners participated in the daily patrols and ambushes conducted by the rifle squads.

The 106 mm recoilless rifle and 81 mm mortar tube were crew-served weapons too large to take on patrols. Therefore, they remained on Hill 41, from which they could provide supporting fire to the patrols and counter battery fire against enemy rocket sites. That, of course, was a great benefit to me as a 106 crewman. Instead of having to spend my nights on patrol, walking miles and sitting in ambush positions, I would remain with the 106 at its fixed position on Hill 41.

Hill 41 was a small place, so it wasn't hard to locate the 106 recoilless rifle team in a tent near the front gate. I entered the tent and introduced myself. The senior person was a Buck Sergeant. I was welcomed by the team and assigned an open bunk. The dozen or so members of the squad who shared the tent slept on government-issue canvas folding cots in one big open area. Cabinets made from ammunition crates held personal gear and were adorned with pinup girls and calendars with months and days marked off.

My new colleagues in the 106 team had names like Panda Bear, Opie, Wisenheimer, and Moose. Like most other Marine units, they came from all over the United States. Some had only been in Vietnam for a few weeks, some were in mid-tour, and one would be rotating home soon.

The Shower

The living conditions on Hill 41 were not bad for being in the Marine Corps infantry in Vietnam. I certainly had it better than the 0311 riflemen. One thing that I missed, however, was the ability to take a hot shower. The shower consisted of a 55-gallon oil drum sitting on a framework constructed of 2x4 boards. You stood under the drum and reached up to hold open the valve to release water from the drum that would trickle onto your head and run down your body. To fill the drum with water, it was necessary to climb a ladder and pour water from a bucket into the drum. The water had to be transported from a water supply off the hill. The only water on the hill was in a large tank on wheels, which the Marines called a "Water Buffalo" and could only be used for drinking.

Determined to find a way to have a hot shower, I began to experiment. My first method was to drip gasoline into a submerged chamber where it would be ignited by a flame, much like the devices we used to heat water to clean our trays at the mess hall. I was behind the "hootch," dripping gasoline from a C ration can into a bucket with flames in it, when I was distracted and allowed the drip to become a stream. Fire from the bucket ran up the stream toward the gasoline-filled container that I was holding. Seeing the flame running up the stream of gasoline, I threw the can, which caused a fireball of flaming gasoline, some of which covered my hands and arms. I ran into the "hootch" flailing both arms to put out the flames.

The guys in the tent looked up from their card game in amazement. The gasoline burned off without causing any permanent damage. Someone referred to me as "Genius." I don't think it was meant to be flattering. Any negative opinions were overcome when, a few days later, when I perfected the hot water heater and the residents of our "hootch" were standing in line for hot showers. The name, however, stuck.

The 106 team on Hill 41 now included a Lance Corporal known as "Genius."

THE MISSION

The mission of the 7th Marines was driven by a new and serious threat. The North Vietnamese had begun to use long range rocket artillery against major military installations, including the airbase and military complex at Da Nang.

Soldiers of the North Vietnamese Army, along with the laborers necessary for the task, had been transporting their heavy loads from North Vietnam. Trucks were used at first along the system of trails and roads known as the Ho Chi Minh Trail, leading south inside Laos along the border with South Vietnam. The trail was a clandestine superhighway along which most of the personnel and supplies necessary to live and fight were transported. It was safe to move along most of the trail because much of it was cut through triple canopy jungle, which blocked prying American and South Vietnamese eyes. And of course, it was in Laos.

When the burdened NVA unit arrived at a point to the

west of Da Nang, they left the main part of the trail and turned east through the mountains. The laborers carried their heavy loads eastward along paths cut through the jungle. The pathways resembled tunnels cut through foliage. The tunnels were just high enough to allow the Vietnamese, but not average Americans, to move freely. I have moved along those trails and found it necessary to bend at the waist to move through them.

Eventually the mountain trails ended and the party came to a valley. The flat valley floor was covered with elephant grass and led to villages with cultivated rice fields. The journey was becoming more dangerous. The NVA group was now inside South Vietnam and, if detected, would be immediately attacked by American forces who would have high technology equipment, artillery, aircraft, and no mercy. They would move only at night for the few miles remaining in their journey. Only the top commanders of the NVA unit knew their destination, their target, or the time of their attack.

After dark, the party ventured out of the jungle into the flat land known by the Americans as "Happy Valley." A short distance away was a river where local Viet Cong with sampans waited. The bearers were relieved to put their heavy cargo into the boats and ride along with them until they came to a village comprised of several hamlets, known as the Bo Bans, near the convergence of the Tuy Loan, Yen, and Can Do Rivers.

There, they unloaded 130 Soviet BM-14 140mm rockets. Each rocket was 42.3 inches long and weighed 90 pounds. Using shovels and other digging tools, they prepared launch positions for each of the rockets, 10 meters apart. Each rocket was positioned using aiming stakes set by NVA rocket engineers. The engineers had used theodolites and transits to determine the proper azimuth and elevation that would cause the rocket and its explosive charge to impact the sprawling American military complex and huge airbase in Da Nang. Electronic ignition wiring was attached to each rocket.

When the NVA commander gave the signal to launch, 66 rockets whooshed into the air, leaving bright flashing trails of fire almost 1000 feet behind them as the propellant provided thrust to take the rocket and its payload the several miles to Da Nang. While 64 of the rockets had failed to fire, the others reached their target. At 0310 on the morning of 27 February 1967, enemy rockets fell on the Da Nang airfield for the first time. Eleven Airmen and one Marine were killed, 35 Americans were wounded, and 13 aircraft were damaged.

Not all of the rockets hit the American airbase. The rockets killed more Vietnamese civilians than Americans: 35 villagers died in the attack and 70 were injured. Between 150 and 200 Vietnamese homes were destroyed. The NVA, thanks to the help of the Soviets, had a devastating new weapon. The war had changed. The mission of the 7th Marines would be tailored to meet this new threat.

Working in The Rocket Belt

At the time of the first rocket attack, "Happy Valley," southwest of Da Nang, was in the area of tactical responsibility of the 9th Marines. The mission then was to patrol the area to prevent enemy forces, especially sappers, from reaching the US installations around Da Nang. Stopping the enemy in the mountains was impractical because the jungle terrain provided them cover and concealed their movements. Between the mountains and Da Nang was the open land in the valley. In the unpopulated areas of the valley, there were areas where elephant grass grew as high as 10 feet, providing concealment. The closer to Da Nang, the more open and visible the area became. Rice fields and villages were scattered along the valley, especially near the rivers.

From the time of the first rocket attack in February 1967, a band of territory circling Da Nang was known as the "rocket belt." It was no longer sufficient to keep sappers and NVA infantry from transiting the area to approach Da Nang. It was now important to prevent the enemy from entering the area, staging, preparing, and launching rockets.

The Marines executed that mission by patrolling the area both day and night. Day patrols looked for signs of enemy movement, launch sites, or stashed rockets. Before a rocket attack, the enemy would bring the rockets out of the mountains and into the valley where they would be hidden until the day of launch, two to 30 days away. Local Viet Cong, who would smile politely at Marine patrols from their rice fields or houses during the day, would assist the NVA by preparing hiding places for both the rockets and the NVA in and around their villages. Systems of tunnels, caves, and other hiding places were constructed and maintained by local Viet Cong. Hundreds of enemy troops could be hidden in and around a small village.

Most of the activity was at night. On the afternoon before the launch, the enemy engineers, perhaps dressed as peasants and carrying visible farm implements, would visit the launch site and use their specialized equipment to measure the azimuth and elevation appropriate to the target. The engineers would then place aiming stakes to assist the rocket crews in placing the rockets in the proper position.

After nightfall, the rockets would be removed from their hiding places and transported to the launch site. Rockets were staged where they could be transported to their launch site, over land or water, within an hour and a half. Preparing the rockets to launch would take up to three hours. A Viet Cong defector stated that the usual setup time for 122mm rockets was an hour and 15 minutes.

Sometimes, the transport or setup procedures would be disrupted when they were discovered by Marine patrols, most often during the transit stage. Marine night patrols were actually ambushes. The patrol would leave their combat base during daylight hours. After dark, the Marines would conceal themselves in a position from which they could watch trails or open areas where infiltrating enemy would likely be moving. The patrol might remain in one place or move to two or three different positions during the night.

Of course, there were not enough Marine patrols to cover all the possible paths, trails, villages and potential launch sites in Happy Valley. The first mission was to prevent rocket attacks by keeping the enemy, and his rockets, out of range of Da Nang. The second mission was to stop rocket attacks in progress and kill the rocketeers. That was my job.

LARRY E SMEDLEY MEDAL OF HONOR

The squad leader looked out over Happy Valley. He did not like what he saw, or more accurately, what he could not see. In the dark and moonless night, he could not see well enough to detect the movement of enemy troops as they moved through the valley below the knoll on which the Marine rifle squad stood. The quarter moon would provide a luminosity of 92 percent, which would result in excellent moonlight. However, the moon would not rise until 2120 hours and would disappear over the horizon again before midnight, draping the valley in darkness once again. For the greater part of the night, the visibility of the valley below would be insufficient for the mission.

The knoll was the position where his squad had been assigned to set up a night ambush. Enemy movement was expected in this area tonight. All three squads from the 1st Platoon of Delta Company, 1st Battalion, 7th Marine Regiment, had been assigned positions across the mouth of the valley, from which they would be able to see any movement along the valley floor. Intelligence reports indicated that the North Vietnamese Army (NVA) and Viet Cong (VC) were expected to move into the area this night to launch their 122mm rockets toward the vital Da Nang military complex and airbase.

The Marines were entrusted with the enormous responsibility of preventing attacks that could kill large numbers of Americans and their allies and destroy millions of dollars worth of aircraft and other equipment. Every night, squad size patrols left the combat base of Delta Company to patrol the valley and set up ambush positions with orders to prevent the enemy from launching rockets from the area known as the "rocket belt." A Marine rifle squad should be comprised of three fire teams of four men each, plus the squad leader, who would hold the rank of Sergeant and have significant experience. The squad leader had the duty of deploying his men for the best tactical advantage and directing them during chaotic and frightening firefights. He was given the authority to order that deadly force be used, and he could call for artillery support or medical evacuation helicopters if needed. There should also be a Navy Corpsman to render medical aid in case of injury to a squad member. Finally, a machine gunner and assistant gunner, assigned to the weapons platoon of the Company, would be attached to the squad.

The past several months had been difficult for Delta Company. Combat attrition had severely degraded manpower. There were only six men in the squad overlooking the valley. There was no corpsman. There was no machine gun. The Marines, most of them teenagers, were armed with six M16 automatic rifles and one M79 "blooper" 40mm grenade launcher. Although not yet old enough to vote, drink beer, or own a firearm in many US states, they were sent into one of the most dangerous environments on the planet with instructions to kill anyone they found there. The leader of this small band of teenage Marines was Corporal Larry E. Smedley, age 18, from Orlando, Florida.

Born in Front Royal, Virginia, Smedley had moved to Orlando where his father was stationed at McCoy Air Force base. His mother worked at a Walgreens drug store. Just after turning 17, Smedley quit Colonial High School and joined the Marine Corps. At that age, it was necessary to have the permission of parents or guardians to join the military. With the blessings of his parents, Smedley was off to Parris Island, South Carolina, for boot camp. Less than two years later, he was leading a Marine night ambush patrol in Vietnam, making tactical assessments, giving orders to Marines in combat, and making life or death decisions.

The patrol's ambush position had been selected by the Platoon Commander, 2nd Lieutenant Michael Neil. Smedley was not free to change it without Lieutenant Neil's authority. It was important for the Platoon Commander and the command center on Hill 41 to know the position of all the Marine elements in the area. Allowing the squads to roam freely through the dark, ready to shoot anything that moved, invited friendly fire incidents. Smedley would have to get permission from Lieutenant Neil to move his ambush location.

Lieutenant Neil, at the age of 27, was much older than his contemporaries. A native of Long Beach, California, Neil graduated from San Diego State University before attending Boalt Hall School of Law at the University of California, Berkeley, and graduating in 1966. Berkeley was home to radical students and faculty who openly supported the Viet Cong and North Vietnamese and opposed their own country. The treasonous conduct of Berkeley students and faculty was blatant. There wasn't even a colorable attempt to pass the actions off as "anti-war." A January 1966 article in the Madera Tribune newspaper reported that a committee of students, led by a former student, collected funds for the purpose of supporting the Viet Cong and forwarded that money through front organizations in Prague and Geneva to the people that Neil was now fighting.

Neil was not your typical Berkeley student. One day he attended a movie on campus. The film showed an American helicopter on the ground in Vietnam after being shot down by the Viet Cong. The young blonde pilot was pulled from the wreckage and killed on screen. The Berkeley students cheered. Neil walked out. Disgusted by the behavior of the Berkley students, he joined the Marines.

Smedley used the squad radio to contact Lieutenant Neil, explaining that visibility into the valley was insufficient to detect movement and requesting permission to move his ambush position from the knoll to the valley floor. Smedley told Neil that he thought he could "get some VC" from his proposed ambush site. Neil agreed and granted the request. There was a tactical disadvantage to being on the valley floor. When the NVA and Viet Cong launched a rocket attack, they used a large force. A 122mm rocket weighed 121 pounds and took two men to carry, even when it was broken down into its three component parts. There would also be infantry troops, engineers, and other specialists to aim and fire the rockets. It was possible that the approaching enemy force would walk right into the Marine position. From higher ground, such as the first position on the knoll, the Marine squad could observe the approach of an enemy force from a safer position, report the movement of the enemy, and call in artillery. Nobody would have faulted the young squad leader if he had followed his orders and remained on the knoll.

Approach of the Enemy and Initial Contact

The spot Smedley chose as his new ambush location was a small thicket near the village of Phouc Ninh (2). The Marines settled into the thicket to wait and observe. About an hour later, the shadows in front of their position began to move. A large enemy force emerged from the dark night, moving directly toward the thicket where the Marines remained motionless, with racing pulses and building adrenaline levels. Smedley whispered into the radio handset that he had movement in front of his position and estimated that there were 100 "vermin," a code for enemy troops. As the Marines watched breathlessly from their concealed position, the enemy force passed within a few feet of them, on both sides of the thicket. At one point the enemy column stopped, right next to the thicket. The Marine radioman found himself standing so close to the enemy radioman that he could almost reach out and touch him. The Marines were hidden by the wooded thicket and the moon shadow that it provided. The enemy, looking toward the thicket, saw only darkness and shadows.

What could give the Marines away now would be movement or sound. The Marines made neither a whisper or movement. Movement caused noise. Even a little noise could give away their position. They all had black electrical tape around their two dog tags that swung on chains from their necks to keep the tags from producing a "tinkle" sound when they moved None wore cologne, aftershave, or scented deodorant. These were standard precautions taken by Marine infantrymen who hoped to never be close enough to the enemy to have these precautions matter but were grateful now for having taken them.

The Marines, looking into the moonlight from the dark thicket, had a good view of the enemy soldiers. After a few seconds that seemed like a hundred years, the enemy column moved forward again. The Marine radioman held up the handset of the radio to Smedley, who moved his head from side to side, declining the radio. There would be time to use it after the "vermin" had passed. The enemy soldiers were carrying mortar tubes, machine guns, and objects that looked like rockets as they moved in the direction of Hill 41. The Marines continued to watch for several minutes as a steady stream of enemy troops filed past, only a few feet away.

The rockets were obviously intended for the sprawling Da Nang complex and airbase. A common practice of the NVA and VC rocketeers was to attack any nearby US position with mortars to prevent counter fire against the rocket launch site, and Hill 41 would have been close to that site. The most effective weapon against rockets was the 106mm recoilless rifle. There was a 106 on Hill 41, and it was always manned and ready to fire during the night hours when rocket attacks were likely. Within about 10 seconds of being given permission to fire, the recoilless rifle could fire a large and deadly high-explosive warhead into a rocket site as far as 3,500 meters away. The tactic of suppressing counter-fire from the 106 by attacking its position with mortars never worked, but the Viet Cong continued to try.

The mission of the Marines was to patrol Happy Valley in order to detect enemy forces attempting to reach or fire rockets toward the valuable complex in Da Nang. The Marines were ordered to engage any enemy force they encountered. The fact that the Marines were outnumbered about 15 to one did not create an exception to those orders.

Having been alerted by Smedley's first radio call that a large enemy force with rockets was moving into position to launch an attack, Lieutenant Neil began to move other assets into positions to intercept the enemy force and support Smedley and his squad. The two other squads from 1st platoon were in ambush positions in the area. Lieutenant Neil was with the squad led by Corporal Matthew Petit. The third squad was led by Corporal Norman Kurtz. Both squads moved as quickly as tactically possible toward positions ahead of the enemy to block their path. On Hill 41, Lieutenant Jack Perkins was scrambling two squads of Marines onto Amphibious Tractors (AMTRAC's) and preparing to send them as a reactionary force to assist the three squads of 1st platoon.

The Marines moving to block the enemy path and reinforce Smedley's squad would not make it there in time to coordinate an attack. The enemy was slipping away into the night, still unaware that they had been detected and that five squads of Marines would soon be engaging them. After the last enemy soldier had passed the thicket, Smedley used his radio to contact Lieutenant Neil, reporting what he had seen and expressing his intent to attack the enemy force. Neil gave his approval. The Marines slipped out of the thicket, lined up side by side, and opened fire. Red and green tracers lit up the sky as astounded and confused enemy rocketeers fled what they must have assumed to be a large and stealthy American force. It is not known what casualties were inflicted upon the enemy at that point, but Da Nang would not be attacked that night, not by this rocket force.

The enemy rocketeers knew there was no chance to successfully complete their assigned mission. It takes up to two hours to set up the rockets and prepare them for launch. The Marines would catch up to them in far less time than that. They found themselves in a difficult position. Usually, they fled swiftly from their launch site after the last rocket had been fired, carrying only the launchers which, although awkward, could be broken down into component parts and shared among several men. Tonight, they had to carry the rockets as well. They could not move as fast as the pursuing Marines. To their advantage, they had broken contact with the Americans, who seemed to have lost their trail.

In fact, Lieutenant Neil had directed Corporal Smedley to hold up on his pursuit of the fleeing enemy force until the platoon commander and the squad led by Petit could link up with Smedley. Having three uncoordinated squads moving in the same hostile area in search of the enemy presented dangers. More than once, Marine squads had exchanged fire with one another in similar circumstances.

Rockets

A North Vietnamese Army rocket battalion is equipped with 18 launchers. Because the rocketeers will leave their launch site within 15 minutes to avoid the response of US forces, which is certain to come, each launcher will have time to fire only three rockets. Therefore, a battalion-sized rocket team will launch 54 rockets. Each NVA rocket battalion is also equipped with two machine guns that will protect the perimeter of the rocket site. This means that 108 troops are necessary just to carry the rockets. If rocket launching tubes are used, the size of the force necessary is increased by four to six men for each of the 18 tubes, which weigh 110 pounds. Therefore, Smedley's estimate of 100 enemy troops was likely a significant underestimate.

Linkup

The other two squads of the 1st platoon left their ambush positions and moved to support Smedley and his men. The platoon commander, Lieutenant Neil, and his radio operator, Ken Dohanish, were with Petit's squad. When all three rifle squads of the 1st platoon linked up, Lieutenant Neil directed his platoon to sweep the area where the enemy rocket force was last seen. The combined platoon included a Navy

Corpsman and a machine gun team, which had been assigned to one of the squads.

To gain better visibility during the hunt for the enemy force, the Marines requested illumination. Soon, the familiar "pops" were heard overhead as artillery parachute flares were ejected from artillery rounds and ignited, slowly descending under parachutes and bathing the area in their eerie yellow light. Each time a flare was ejected from its artillery round casing, the "whoosh-whoosh-whoosh" sound of the spinning casing falling away was heard, followed by the sound of the metal casing crashing into the foliage. An airplane soon appeared overhead and began dropping aerial flares six feet long that made the area look like it was mid-day. Illumination is a good tactical tool, but it has the same disadvantage as tracer rounds. They work both ways. The flares, swinging from their parachutes and causing shadows to sway back in forth, illuminated whatever their light fell upon, including the Marines.

Lieutenant Neil led the Marines in a sweep of the area in search of signs of the enemy. About 400 meters into the sweep, the Marines had just topped a small rise, where they were silhouetted by the light of the flares behind them, exposing them visually. Even the least experienced Marine rifleman knew that, to the greatest extent possible, it was best to stay off ridge lines and hilltops, even in the daylight. On a ridge line, with only the sky behind you, you stood out. Once you came off the ridge line, you blended into the background. Being on a hilltop or ridgeline, unless in a fixed concealed position, makes bush-savvy Marines nervous and extra cautious.

• • •

Second Action

As they began to descend the backside of the rise and were on their way toward the bottom in two columns, they heard a "swoosh." It took a second to realize that the sound came from a rocket-propelled grenade (RPG) that had just shot past them, directly between the two columns. The RPG impacted the ground and exploded near the end of the columns, sending its shrapnel flying. Luckily, the only damage caused by the RPG shrapnel was to the last Marine in the column, who suffered a small cut on his lip.

Suddenly, a large volume of small-arms fire and rocket-propelled grenades erupted from three sides. Lieutenant Neil immediately ordered the platoon to charge forward toward the muzzle flashes at the bottom of the hill, directly into the enemy ambush in typical Marine style. Lance Corporal Ralph Bobian, an 18-year-old machine gunner from Denver, Colorado, aggressively assaulted the enemy position, firing his M60 machine gun from the hip. When a Marine nearby was hit and fell to the ground, Bobian placed himself between the casualty and the enemy machine gun. While still firing his machine gun from the hip, he was struck by machine gun fire and fatally wounded. An enemy machine gun poured its fire into the rushing Marine line. One Marine was startled when something he could not identify struck his flak jacket and left arm. It was the blood and flesh of the Marine next to him.

Corporal Smedley was struck in the upper body with machine gun fire, in addition to having a piece of his foot blown off by a rocket-propelled grenade (RPG). One of the Marines took out the enemy machine gun with a light anti-tank weapon (LAW), a bazooka-type rocket fired from a disposable tube. Almost immediately, the Marines came under fire from a second machine gun concealed in the foliage at the bottom of the hill.

Frank Kuykendall saw the second machine gun just before it opened fire. Kuykendall was only a few feet away, and the enemy gunner was waiting until more Marines came into his field of fire before opening fire. Kuykendall knew that he would be shot by the machine gun, because it was aimed directly at him. A second after he saw the gun, it opened fire, with bullets hitting him in the torso and in his right hand.

Knocked to the ground by the impact of the machine gun bullets and bleeding from his wounds, Kuykendall looked up to see the machine gun still aimed at him at close range. He tried to grab his rifle with a hand that had been rendered useless by the bullet that passed through it. His hand would not close around the grip of his rifle. He knew that the next burst of fire from the machine gun would kill him. Just then, the machine gun swung away from him, moving to a target on his left.

Kuykendall's gaze followed the direction of the machine gun barrel and saw Larry Smedley hobbling by on his wounded foot, moving with determination, straight toward the machine gun. In his hand, he held a grenade. The machine gun opened fire on Smedley, striking him in the chest. Smedley did not slow or falter; he continued directly toward the machine gun, even as it poured a steady stream of fire at him. Smedley dropped his grenade into the machine gun position, killing the gunner and ending the significant threat to his fellow Marines.

Smedley fell to the ground in front of the disabled machine gun position. Lieutenant Neil was armed with only his model 1911 Colt .45 automatic pistol. Arriving at the bottom of the hill, Neil removed his own flak jacket and placed it over Smedley to provide further protection for the wounded Marine. Neil picked up a wounded Marine's M79 grenade launcher and began to fire grenades into the enemy position. Small arms fire and RPG's continued from the woods at the bottom of the small hill. Corporal Norman Kurtz, the leader of one of the three squads of 1st Platoon, was also hit by rifle fire while assaulting into the enemy position.

Lieutenant Neil was using the PRC 25 radio to contact the Delta Company command center on Hill 41 to request medical evacuation (MEDEVAC) for his wounded. His voice, shouting loud to be heard over the sounds of battle, attracted the attention of enemy soldiers who had moved around behind the Marines. Neil and his radio man, Ken Dohanish, began to take fire from an enemy position less than 40 meters away, halfway up the rise. Anyone using a radio was most likely a leader and therefore a target. Killing the radio man, especially if the radio could be disabled by fire at the same time, would provide a tactical advantage. Depriving the Marines of communication ability would prevent them from calling in supporting fire from artillery or aircraft or requesting assistance from a reactionary force.

Ambush from Behind

Lieutenant Neil directed the only squad leader left alive

and unwounded, Lance Corporal Matt Petit, to "kill those guys," motioning toward the direction from which the fire came. Petit selected two of his three-man fire teams and quickly huddled with them to convey his plan. He directed the Marines to fix bayonets and move up the hill toward the enemy soldiers. When ordered, the two fire teams were to attack the enemy position in a series of "fire team rushes." The three-man fire teams would take turns rushing toward the enemy position and firing their rifles into the area where they thought the soldier was hiding. The non-moving fire team would cover the rushing fire team.

Petit observed from a position off to the side of the assaulting fire teams. As the teams were advancing on the enemy position, fire team leader Dave McMurray shouted, "Matt, one is headed right to you." At that moment, Petit saw a dark-haired figure running toward him in the darkness, flushed from his hiding place by the rushing fire teams. The enemy soldier was closing fast when Petit raised his rifle and fired several rounds into him, knocking him to the ground. Suddenly, the NVA soldier rolled over onto his side, moving his hands toward his body. McMurry exclaimed, "He's going for a grenade." Before the enemy could get his hands on the grenade attached to his equipment, Petit thrust a bayonet into him.

After the second engagement

After the Marines secured their position at the bottom of the hill where the enemy machine guns had been, Lieutenant Neil assessed his situation. Two Marines, Smedley and

Bobian, were dead. Five others were wounded badly enough to need evacuation. Back on Hill 41, 1st Lieutenant Jack Perkins, the Executive Officer of Delta Company, had almost completed preparations to send a reactionary force of two additional squads to reinforce the 1st Platoon. Soon, helicopter rotors could be heard overhead, but no aircraft could be seen in the night sky. The helicopter displayed no lights.

The helicopter crews knew that they were close to the platoon who needed help but did not know the platoon's exact location. A complicating factor was that the enemy force was very close to the Marines and therefore very close to any helicopter that came to their assistance.

Marines often needed to call for assistance from a helicopter, usually for a medical evacuation. The general location would be provided to the crew by grid coordinates. As the helicopter got closer, visual signals were used. During the day, smoke grenades not only showed the helicopter crew where to find the Marines, the smoke was helpful to the pilots in determining the direction and speed of the wind in the landing area. At night, aerial flares were used.

Both the smoke signals and flares came in different colors. Dohanish, the radio operator, was talking with the helicopter crew, but neither the crew nor the Marines could see the other. When the helicopter was heard overhead, Dohanish told the crew to expect a flare. The corpsman fired a red flare. So did the NVA! The Marines told the helicopter to expect another flare. Immediately after the corpsman popped a green flare, a green flare shot up from the enemy position! The helicopter crew could not tell the enemy position from the Marines location. To complicate matters, every time the Marines fired a flare, they drew heavy fire.

Unable to establish visual contact by the use of the flares, the Marines attempted to display their position to the helicopter by the use of a white strobe light that they carried for that purpose. Both the corpsman who was treating the wounded and Ken Dohanish, the radio operator, tried to use the strobe light, but each time they turned it on they came under intense fire. The five wounded Marines would have to wait for daybreak to be evacuated.

The closeness of the enemy made it difficult for a helicopter to safely evacuate the wounded and impossible to call in artillery fire without putting themselves in as much danger as the enemy. Artillery was being fired, however, into the area between Phouc Ninh (2) and the mountains, to which the NVA were thought to be fleeing.

The NVA faced a difficult situation. They had been detected and attacked before they could set up and fire their rockets. A number of them had been killed. Their two machine guns had been destroyed. The Marines knew their location at this moment. They needed to get out of the area and into a hiding place before dawn. The sun would rise this Thursday morning, December 21, 1967, at 0710. First light would be about 0640. The NVA needed to move themselves, their equipment, and their wounded to a safe place before first light. They slipped away, taking their rockets and wounded, but leaving two rocket launchers and the bodies of several of their comrades. Unknown to the Marines, an NVA soldier who had become separated from his unit during the fight was hiding in bushes about 10 feet from Lieutenant Neil and Dohanish. Realizing that he would soon be discovered as the dawn light deprived him of the cover of darkness, he stood up with his rifle and was immediately shot dead.

With the dawn came AMTRACs with two squads from Hill 41. Helicopters arrived to take out the casualties. The Marines on the AMTRACs had arrived in the area earlier but did not approach during the darkness and confusion of the firefight.

Discovery of Enemy Bodies and Rocket Tubes

As soon as there was enough light to do so, the Marines searched the area. In addition to several dead bodies in NVA uniforms, weapons, and other items commonly found after a battle, the Marines found a coveted prize. Not one prize, but two. The NVA were using a Russian-made launcher for their 122mm rockets. Previously, they had used crude aiming devices constructed from bamboo stakes or simple mounds of dirt. That method was good enough to have the rocket land somewhere in the Da Nang vital area, simply because that area was too large to miss. The Russian-made launcher, with its long metal tube and sophisticated optical sight, allowed much more precise targeting. From prisoner interrogation and informant information, US intelligence knew the launchers existed, but no western eyes had ever seen one. Units in the field were instructed to keep an eye out for one of these devices and make every effort to capture one. In their desperation and haste to escape, the NVA had left two of the rocket launchers, including the optical sight for one of them.

There were seven dead soldiers in NVA uniforms scattered around the area, five of them near the rocket launchers.

Bodies and equipment brought back to Hill 41

When the AMTRACs returned to Hill 41 with the Marines from the firefight and everything they had recovered, those of us on the hill gathered around to gawk at the enemy equipment. Several dead NVA soldiers were laid side by side outside my hooch. A rocket tube had been assembled by Marines who had managed to work out how its components went together. My inspection of the enemy gear was interrupted when I was directed to go with another Marine to the helicopter landing zone at the top of the hill and move the two Marine bodies to an empty ammunition bunker nearby.

Two green body bags lay beside each other at the edge of the landing zone. The other Marine and I did as we were told, moving the bodies into the ammo bunker, where they would be out of the sun and elements. That seemed to be the most respectful way to treat our fellow Marines. It was not the first or last time the little ammo bunker would be used for that purpose. It seemed to always be empty of ammo and available for the temporary storage of our dead. Perhaps it was called the ammo bunker because calling it our "Morgue" did not seem appropriate. The Marine bodies were only temporarily stored there until a helicopter could pick them up and take them to the very busy facility in Da Nang, where our dead were prepared for their trip home.

• • •

More NVA bodies found

It was not a surprise to learn that the enemy force suffered more casualties than the Marines had located during the sweep of the area at dawn on December 21. In addition to the seven bodies found at the scene of the fight, there was a significant amount of blood, indicating that many of the NVA soldiers had suffered serious wounds. Patrols in the area over the next two days located fresh graves in the wooded areas near the villages. The patrols were ordered to dig up the graves and take the clothing off the occupants, who had been there for two days now. Intelligence wanted the uniforms to help identify which North Vietnamese Army units were involved in the attempted attack.

Marines in the bush carried small folding shovels known as entrenching tools, or in Marine-speak, "E tools." However, on day patrols near the combat base, there was no need to dig a fighting hole, so the E tools were not carried. This means the Marines had to dig up the graves using their hands, or anything they could find lying around that would fit the purpose. One Marine used the empty canister of an artillery illumination round found nearby. Many of the bodies had been wrapped in the parachutes that littered the area around the battle site. The parachutes had been attached to the large illumination flares dropped by aircraft. The bodies were in such bad shape that some of the Marines vomited at the sight and smell as they exhumed the decomposing enemy soldiers and removed the clothing.

Awards

The Meritorious Unit Citation was awarded to individual units, usually regiments, battalions, and sometimes companies that performed exceptionally. During the entire Vietnam War, only one Marine platoon received that award. It was the first platoon of Delta Company, 1st Battalion, 7th Marine Regiment, 1st Marine Division, as a result of the actions of that platoon on the night of December 20, 1967.

The members of that platoon prevented a rocket attack on Da Nang by attacking and pursuing an enemy force that was vastly superior in number. They also fulfilled an intelligence requirement to capture the first ever 122mm Russian-made rocket launcher.

The Marine Corps is conservative in recommending and presenting awards. In addition to the unit recognition, the individual awards that resulted from that action included the nation's highest award for bravery, the Medal of Honor, to Larry Eugene Smedley, the 18-year-old squad leader who led his six-man squad in an attack on an enemy force of at least 100 troops and sacrificed his life taking out a machine gun that was shooting and killing members of his squad.

Lieutenant Neil was awarded the Navy Cross. Corporal Norman Kurtz and Lance Corporal Matthew Petit, the other two squad leaders of the platoon, each received the Silver Star Medal, as did the machine gunner, Bobian, who died in the shootout with the first machine gun. Other platoon members, including Corporal Ken Dohanish, received the Bronze Star Medal.

Most importantly, nobody in Da Nang was killed or injured in the rocket attack that did not occur on the night of December 20, 1967.

• • •

Smedley honored by his home town

Fifteen years after the fight at Phouc Ninh (2), I was an intelligence officer in the U.S. Navy Reserve. While doing my annual active duty in Washington D.C., I went into the book store in the Pentagon. There, I noticed a small paperback book, "History of the Seventh Marines." I thumbed through the short book, scanning the contents. There was a section about the members of the Seventh Marines who had been awarded the Medal of Honor. A young Marine in dress blues stared at me from the picture taking up the top half of the page. I began to read the citation:

While serving as a squad leader with Company D, 1st Battalion, Seventh

Marines, …

I noted that I, too, had been assigned to Delta Company.

on the evenings of 20-21 December 1967, …

Now I became very interested. I was with Delta on that date.

an estimated 100 Viet Cong and North Vietnamese Army regulars,

carrying 122mm rocket launchers and mortars, were observed moving

toward Hill 41 …

I realized that I was on Hill 41 on December 20, 1967! I wondered if I knew the Marine in the picture. As I continued to read the Medal of Honor citation of Corporal Larry Eugene Smedley, the incident came back into my memory.

There is no time to dwell upon memories in a combat

environment. Every day is a new day, with a new opportunity to die unless you keep focused on your mission. What happened yesterday is history. That particular event was memorable because a Colonel flew out from Da Nang to look at the captured rocket launcher and take it back to be examined for intelligence value. I had never seen a Marine Colonel or a 122mm rocket launcher until that day. As I thought more about it, I did remember Corporal Larry Smedley. I had visited his hootch on Hill 41 to say hello to a fellow Floridian and had joined members of his platoon listening to him play his guitar. I took the fact that a member of Delta Company had been awarded the Medal of Honor in a battle that I remembered and filed it in the back of my mind in the "cool to know" category.

A year later, I attended a ceremony at the Orange County Convention Center in Orlando, Florida. It was the dedication of a memorial to service members from Orange County who were killed in Vietnam. The names of the fallen were engraved on the memorial, which had a fountain at the base. We were all given a flower to put in the water in the fountain. I stood with my flower, reading the names on the memorial. One name staring at me from the gray granite caused me to gasp audibly. LARRY E. SMEDLEY. My mind went into overdrive … wait … could that be the Larry Smedley with whom I had served in Vietnam? The one who was awarded the Medal of Honor? Whose body I put into a sandbagged bunker on Hill 41?

Nobody in Orlando seemed to know that our town had a Medal of Honor recipient. I resolved to change that. Before the internet, research was difficult and tedious. I set about finding Smedley's family and his grave site. I also sought to bring some recognition to the 18-year-old Marine. I was a County Judge at the time and restricted by the rules of judicial conduct from asking people to do things or contribute value. However, I knew people who could, and who were well-regarded in the community. Charlie Reese, a columnist for the Orlando Sentinel newspaper, and Lawson Lamar, the State Attorney, were both eager to help honor Larry Smedley. Lawson had been an Army intelligence officer in Vietnam. We formed a small committee along with Colonel Joe Kittinger, a retired Air Force officer and Vietnam veteran, as well as several other local leaders. Charlie wrote an article for the Sentinel, informing our community that a young man from Orlando had been awarded the nation's highest award for bravery and suggesting that he should be honored. Charlie provided the address where readers could send donations to the committee for that purpose.

The committee raised enough money to put together a simple but impressive memorial. A local concrete company provided the base of a stand for a large bronze plaque setting forth the Medal of Honor citation. The plaque was provided at cost from a local funeral home. The county government passed an ordinance authorizing the placement of the plaque at the Orange County Courthouse in perpetuity.

I located his mother, Mrs. Mary Smith, now remarried and living in Front Royal, Virginia, where Larry was born. I attempted to contact the other members of Smedley's platoon. The platoon commander, 2nd Lieutenant Neil, was now Brigadier General Michael Neil in the Marine Reserve and practicing law in San Diego. Frank Kuykendall was a long-distance truck driver in the Midwest. I reached him by phone the day before the ceremony, with apologies for the short notice. He said, "I'll be there," and he was. He drove all night.

A ceremony was held at the Orange County courthouse to dedicate the memorial to Smedley. The Smedley family came, General Neil traveled from California, the Navy sent a band. In the 20 years since that ceremony, a new courthouse has been built in Orlando with an even larger memorial to Smedley and other citizens of Orange County who have been killed in various wars. A large portrait of Smedley, of outstanding quality painted by Rick Thompson and donated by Lockheed Martin, occupies a place of honor in the rotunda of the courthouse, along with the actual Medal of Honor provided by the Smedley family.

The eighteen-year-old Marine from Orlando had received the honor and recognition he deserved.

HERE THEY COME

In January 1968, the tea leaves were pretty easy to read. The North Vietnamese Army (NVA) and Viet Cong (VC) were planning a massive offensive across the entire Republic of South Vietnam. In the Da Nang area, technical intelligence methods such as airborne personnel detectors, also known as "people sniffers," had detected a build-up of forces in the mountains south of Da Nang and west of the 7th Marines operating area. A reduction in the number of booby traps in the area indicated that the local Viet Cong had removed them to protect enemy troops who might be unfamiliar with the area. Intelligence reports received from several usually reliable sources revealed that significant elements of the North Vietnamese Army and Viet Cong were moving into positions from which attacks could be launched against Da Nang. The second largest city in South Vietnam and the home to a huge US military base would be an obvious target of the North Vietnamese.

The R20 VC Battalion, as well as the 2nd NVA Division, including the 1st VC and 3rd NVA Regiments, reportedly moved into Go Noi Island on the southern edge of the operational area of 1/7.

VC recon

The Viet Cong R20 Doc Lap Battalion had been fighting the Americans south of Da Nang since the war started. They had proved to be a formidable force in the area centered 15 miles south of Da Nang.

When their commanding officer, Major Nguyen Van Lam, received orders to conduct an intelligence gathering mission at the headquarters of the 1st Battalion, 7th Marines, on Hill 10, he knew that the mission would be important to the upcoming major offensive. Therefore, he decided to lead the scouting mission himself. Although many of the locals who worked at the American base were Viet Cong informants, if not Viet Cong fighters by night, the information they had provided needed to be verified.

It was for that reason that on the evening of January 28, Major Van Lam and two of his troops were moving stealthily along a path just west of Hieu Duc, near Hill 10, being careful to avoid the night ambush patrols of the American Marines. The Marines would hide in the dark and wait for Viet Cong to move down the trails, perhaps bringing rockets to fire at the Da Nang base or moving to positions from which to attack the American bases. There were not enough Marine patrols to cover all the trails, but it was important to be vigilant when moving at night.

It was a dangerous business, but Van Lam had been successful at it for years. Until tonight. Suddenly, muzzle flashes in the dark announced the presence of a well-concealed ambush patrol from Charlie Company 1/7, based on Hill 10. Every member of the Marine squad fired at once. Instantly, all three VC were hit with multiple rounds of rifle fire. Although his two companions were killed instantly, Major Van Lam remained alive but seriously wounded.

One of the Americans who had a large medical kit examined the wounds of Major Von Lam while another American searched him for weapons and materials of intelligence value. When the Marine in charge of the patrol discovered the Major's notebook, he spoke to the man with the medical kit, who suddenly seemed to be more motivated to save the wounded prisoner. While the corpsman treated Major Van Lam, the squad leader made a radio call. It was not long before a helicopter arrived to transport the Major to the Naval Support Activity at Marble Mountain, south of Da Nang. American intelligence officers were already on their way there. They wanted to know why a Viet Cong who identified himself as the Commanding Officer of a VC Battalion was personally participating in a night patrol and why his notebook contained detailed sketches of Hill 10, including the command center. It was clear that the Viet Cong had plans for Hill 10.

As the medical staff and the intelligence officers argued about whether to put Van Lam under anesthesia to begin surgery or to keep him conscious to be interrogated, the Viet Cong Major moved on to an ancestral reunion. The R20 Doc Lap Battalion would be needing a new Commanding Officer.

. . .

Recon by 1st Marine Div.

January 29 – mid day

In the Que Son mountains, a Battalion of the 2nd NVA regiment was moving east on foot. They were on their way to Da Nang to join up with other NVA units and elements of the local Viet Cong for the big attack on Da Nang. Several hundred North Vietnamese soldiers were spaced closely together in four columns.

They were late. The attack would begin tonight, but they were still almost 30 miles from Da Nang and would not be able to arrive in time to participate. Some of the troops wondered if their commander had planned to arrive late for a tough battle.

Many of the troops were new conscripts, barely trained and not highly motivated. The unit had hurriedly been assembled and sent off on the long journey to participate in the great offensive in the south. They had been equipped with helmets and flak jackets, which would provide some protection from the weapons of the Americans and South Vietnamese troops they would soon be facing. The down side of the body armor was having to wear it for days while on the march south along jungle trails that rose and fell with the mountain landscape.

After weeks of marching, they would not make it to Da Nang. Suddenly, all throughout the column, explosions sent shards of shrapnel tearing through the column of soldiers. Within seconds, between 50 and 75 well-placed 105mm and 155mm artillery shells landed on the enemy force with devastating effect. As many as 100 soldiers were killed in the initial barrage, which had been coordinated with three different American artillery batteries.

On a ridgeline overlooking the trail, a small team of Marines from a reconnaissance "Stingray" team with the call sign "Rummage" watched the carnage unfold and assisted in its development. The Marines of "Rummage" had been watching the NVA column for some time. They first observed 40 NVA soldiers, who proved to be the forward portion of the advancing force. As they watched through their scope, they saw the main body, which consisted of four columns of soldiers and extended as far as they could see. The recon Marines stopped counting at 500.

"Rummage" had coordinated the artillery fire mission that was now destroying the enemy force. The shelling did not stop with the first 100 rounds; that was just the opening volley. Death and horror in the form of artillery shells continued to rain down on the NVA who were scurrying for, but not finding, shelter from the barrage. The recon team observing from the high ground called information to the artillery batteries, who adjusted their fire to maximize its effectiveness against the confused and stunned NVA.

When the shelling stopped, it was not cause for celebration by the North Vietnamese. It was the beginning of phase two of their destruction. Overhead appeared what looked like a lumbering World War II cargo plane. The plane began to slowly circle in what pilots call a pylon turn. This steep bank maneuver caused the aircraft to pivot around a point on the ground. At that point was an NVA battalion. The AC-47 pilots learned that they could place their fire on a spot between their view of the left propeller hub and the engine cowling as they turned.

The C47 was known as a Dakota when it dropped paratroopers into Europe over 20 years before. This model, designated the AC-47, was known to Marines as "Puff the Magic Dragon" or a later variant, "Spooky." It was a gunship. The term "Dragon" was fitting for the AC-47, because the sound of the mini-guns resembled the roar of a dragon and the red stream of tracers spewing from it was like a stream of fire.

"Puff" was turning left because that is the side of the plane where the three general electric model MSU-470 mini-guns were located. Each gun had the capability of firing 3,000 rounds a minute.

The North Vietnamese troops had been warned about the "Dragon." Captured NVA documents advised troops not to fire at the AC-47 because the NVA weapons could not harm it and the fire would only infuriate the Dragon. This Dragon seemed infuriated to begin with, as the three Gatling guns unleashed a torrent of 7.62mm bullets at a rate of fire that they had only heard about and never seen. Each fifth round fired was a red tracer. The stream of fire coming from the sky looked like a solid red ray of light.

"Rummage" watched as the gunship worked over the enemy soldiers who tried to scatter out of its fire. A group of 50 soldiers tried to get away by crossing a stream. The recon Marines watched as they were trapped in the stream and cut down by an accurate hail of bullets, causing the water to foam and turn red.

The stream of fire from an AC-47 could put a bullet in every square foot of a football field in 60 seconds. At that rate, it did not take long to expend the 54,000 rounds of ammunition on board. By the time the gunship had run out of ammunition and turned toward Da Nang to get more, dead NVA soldiers littered the trail in all directions and floated down the stream. The few who were uninjured administered aid to the many wounded.

Almost instantly, the roar of the Dragon was replaced by the sound of incoming jet attack aircraft carrying fragmentation bombs and the most frightening munition of all, napalm. After several F4 Phantom aircraft completed their bombing runs, the sound of helicopters was added to the range of deadly sounds heard by the diminishing enemy force. Helicopter gunships raked the area, mopping up what was left of an NVA Battalion.

There was no escape for the NVA soldiers. The recon Marines observed their efforts to evade the massacre and relayed that information to whichever platform of destruction was on station at the time. The destruction did not let up until the sun went down, cloaking the area in darkness and hiding what little remained of the force from the eyes of "Rummage."

By nightfall, the reinforced Battalion of North Vietnamese Army was combat ineffective and would play no role in the uprising that was a few hours away.

Other teams

"Rummage" was not the only recon team in a position to observe and report on enemy movement from the mountains toward Da Nang. The 1st Marine Division had deployed eleven "Stingray" teams deep into the territory where the NVA were known to be operating. When the enemy forces moved out of their staging areas, we would know.

Team "Saddle Bag," positioned along the Thu Bon river 20 miles southwest of Da Nang, observed 75 soldiers carrying mortar tubes and called for artillery fire.

About 2,000 meters from "Saddle Bag," team "Air Hose" called for an artillery mission on an enemy force of around 50 troops moving east toward the 7th Marines area of operations. "Air Hose" observed a large secondary explosion indicating a direct hit on ammunition supplies being carried by the enemy. Near the same time and in the same area, "Sailfish" called for artillery support against a force of 200 enemy troops carrying 122mm rockets. The artillery fire was on target. After the artillery barrage, fixed wing and helicopter air assets arrived to bomb and strafe the enemy rocketeers.

In the mountains west of Da Nang, recon team "Ice Bound" watched as an enemy rocket team prepared a firing position and readied their rockets. "Ice Bound" called for artillery support, which resulted in secondary explosions and seven enemies killed in Action (KIA).

It was clear from the observations of the Stingray teams that the enemy was on the march. Enemy forces were coming out of the mountains from several directions and moving toward Da Nang in large numbers. The Marine forces in Da Nang went to 100 percent alert. No Marine in the Da Nang area of operations would sleep this night.

The Initial Attack

January 29 - evening

The recon teams provided advanced warning and intelligence about enemy movement. However, the 11 teams could not cover every approach to Da Nang and became much less effective after dark. Enemy soldiers poured into the area to stage for their attacks.

After midnight, rockets and mortars fell on 15 US and allied bases and positions, including my position on Hill 41. I was startled when two mortar rounds landed in front of the 106 recoilless rifle bunker, where I was on watch. I loaded a high-explosive round into the breech and moved to the gunner's side of the bunker as more mortars landed. Chuck Sikes came up from his observation position inside the bunker and joined me on the gun. We observed mortar tube flashes from two different locations. I fired 20 rounds at the flashes, and the mortar attack stopped. Sometime later, rocket launch flashes were seen, and we fired several rounds at the rocket site. Rockets were falling on the airstrip in Da Nang and the Marine helicopter strip at Marble Mountain, two miles from Da Nang.

A squad from Alpha Company was patrolling to the south of Hill 10 when they observed a group of 10 North Vietnamese soldiers on the south side of the Tuy Loan river, preparing a rocket launch site. The patrol called in artillery and mortars, causing the NVA to flee, leaving behind five 122mm rockets.

Ground Action

January 29-30

Enemy action was not limited to rockets and mortars. The enemy attempted to attack bridges with explosives planted by swimmers. At the main bridge connecting Da Nang to the Tien Sha Peninsula, sentries spotted two swimmers near the main span of the bridge. They killed one and took the other prisoner.

About 0100, a platoon of Golf Company, 2nd Battalion, 7th Marines, observed two enemy soldiers with a wooden box on a raft approaching the Route 1 bridge over the Cu De River, north of Da Nang. The Marines killed them both before they could damage the bridge.

Two hours later, sentries at the Da Nang main bridge observed two VC in the water and several sampans approaching. The Marine sentries took the swimmers and sampans under fire, killing one swimmer, capturing the other, and turning away the sampans. The waterborne attacks on bridges were a total failure.

The enemy ground campaign enjoyed some successes. An enemy sapper team of around 15 men blew a hole in the perimeter wire of the compound shared by the 7th Engineer Battalion and the 7th Communications Battalion inside the Da Nang base. The sappers overran the compound, attacked the command bunker, and threw grenades and satchel charges into the living quarters. Two Marines who had remained in their living area were killed. Marines manning defensive positions fought off the attack, repelling the sappers and killing four of them. Just 30 minutes later, sappers attacked and overran the Marine Division sub-sector Bravo combat operations center on Hill 200, taking out the division radio net until 0400. Four Marines were killed and seven were wounded. The sappers left trails of blood as they faded away.

The fighting continued after daybreak and throughout the next day. The local Viet Cong forces and the R20 Battalion that had attacked Da Nang were now attempting to flee the area. However, they ran into blocking forces of Marine and ARVN units. The fleeing enemy troops were trapped on a small island near the Cam La Bridge and Route 1, near a Marine Combined Action Unit. A fierce two-day battle was fought on the island, ending on the morning of January 31. When the fighting was over, 102 VC were killed and 88 taken prisoner.

7th Marines area of responsibility

January 29-evening

A patrol from Charlie Company 1/7 was about to set into a night ambush position when they were fired upon by an enemy force of approximately 100. Two other Charlie company squad-sized patrols in the area came to the aid of the engaged squad, and about an hour later, a third Charlie company platoon arrived. A firefight ensued that lasted until about 2000, when the enemy withdrew. Charlie company casualties were heavy, 12 KIA and 6 WIA. Three enemy bodies were recovered. The dead NVA were wearing black pajamas under their green uniforms. They seemed prepared to pretend to be civilians to infiltrate the area.

Combined Action Platoons (CAP) units throughout the Area of Operations were being hit hard. CAP units were comprised of a squad of Marines and a local Popular Forces company. The Marines lived in a compound within a village. Popular Forces were the South Vietnamese equivalent of the National Guard in the US. CAP units were relatively easy prey because of the small number of Marines in each, typically eight. On the night of January 31, CAP Bravo-4 observed an enemy force of 600 enemy troops surround Hill 10. CAP Bravo-4 was attacked but repelled the assault.

A lull

During the first five days in February, the fighting slowed down but did not end. The enemy seemed to be regrouping and getting ready for phase two of their offensive. Reports were being received of small units of NVA moving north in the Da Nang area. Intelligence reported that the 2nd NVA Division had shifted its position four miles north of its previously reported location on Go Noi Island. The 21st NVA regiment was now operating from Go Noi. Reports were also received that the 1st VC and the 3rd NVA regiments had moved into the area of Cau Do River, to the south of Da Nang. The NVA was staging for their next offensive.

Enemy presence around Hill 41

The countryside in the area of operations of the 7th Marines was heavily Viet Cong. During the day, the villagers would farm their fields, wave at Marine patrols, and accept with a smile the medical care they received from the Navy Corpsman who stopped by to treat them. At night, they would change into their Viet Cong persona to ambush the

Marine patrols and plant mines on the roads used by US vehicles. The Vietnamese barber on Hill 41 was caught in the act of planting a mine one night.

Concealed cave complexes and underground hiding places were abundant. It was there that the Viet Cong hid their weapons and equipment. Searches of villages and homes rarely turned up evidence of Viet Cong affiliation.

The extensive cave systems allowed the Viet Cong to simply disappear after an engagement with the Marines, taking their dead and wounded with them. The Viet Cong and NVA could remain underground and out of sight for days. Underground hospitals were available for their wounded. Above ground, their supporting sympathizers were tending their fields, watching the movements of Marine patrols and reporting those movements to the underground NVA and Viet Cong soldiers.

After the war, the local villagers would happily describe their cave networks to tourists and brag that their best fighters worked for the Americans during the day and fought them at night.

Enemy activity vicinity of Hill 41

February 3

On February 3, a patrol from Delta Company 1/7 discovered a group of 200 to 300 enemy hiding rockets in caves near Hill 41. On that same day, Sergeant Glenn T. Lunsford was with a group of Marines from Delta Company who discovered another large cave complex near the village of Phu Son (2). Some of the hiding places were very cleverly concealed.

The cave complex at Phu Son (2) could only be entered from underwater through several camouflaged entrances below the river bank. Sergeant Lunsford placed a Marine at each of the entrances and supervised the creation of a new underground entrance to the cave. Once the new opening was created, Lunsford threw in a tear gas grenade. Soon, three Viet Cong soldiers came out throwing grenades. They were shot dead. Sergeant Lunsford entered the cave, armed with a pistol and a flashlight. He moved through a narrow passageway to a large room where he encountered two Viet Cong who were trying to escape through an underwater exit. Lunsford killed one and captured the other.

The captured enemy soldier indicated that several Viet Cong remained in the cave complex. Sergeant Lunsford again entered the tunnel. A Viet Cong soldier hiding around a corner threw a grenade, the fragments of which wounded Lunsford in both legs. Disregarding his painful wounds, Lunsford chased the enemy soldier down and captured him, bringing him out of the cave. Once again, Lunsford entered the cave, along with other Marines who found several weapons and supplies and information of intelligence value.

Getting Closer

Everyone knew that an attack was coming; the question was when. I paid a visit to the ammunition supply at Battalion headquarters on Hill 10 and drew another 40 high-explosive 106 rounds and 3 boxes of spotting rounds.

On the night of February 5, a rocket attack on Hill 10, battalion headquarters of 1/7, resulted in six Marines killed and several wounded. The mess hall, a large tent, was shredded with shrapnel from the 122mm rockets.

Rocket Attack on Da Nang

A waxing crescent moon provided little light as I watched over Happy Valley from the 106 gun bunker on Hill 41. The calmness of the night belied the danger. We were on 100 percent alert. Not a single Marine on the hill was asleep. The night seemed calm. The sun would rise in four hours, and I hoped to get some sleep after daybreak, when the alert would be lifted.

I had turned 19 just a couple of months before. After experiencing life as a Marine in Vietnam, I was just hoping to make it to 20. As a lance corporal, I was pretty low in the food chain and not privy to the intelligence reports that the Company commander and others relied upon. I did know that there had been a surge of enemy activity within the past week. On Feb 1, the previous day, a Combined Action Platoon designated CAP B4 was attacked by a large enemy force.

It was February 2 when, at 0330, the peaceful night sky erupted with streaks of yellow reaching into the sky from the floor of Happy Valley. A rocket attack was being launched against the Da Nang airbase. I was on Hill 41 to deal with precisely this situation. Our entire battalion had the mission of preventing rocket attacks on Da Nang and immediately stopping any attack that could not be prevented. Our night patrols and ambushes were conducted with the purpose of interdicting the enemy gunners who would slip into Happy

Valley from the mountains to the west, bringing with them Russian-made 122mm rockets. There were too many trails to cover with our limited forces. From time to time, the enemy was able to slip in undetected and launch their missiles.

When the prevention part of that mission failed, I went to work. This was what I lived for. I moved quickly into the gunner's position alongside the recoilless rifle. There were two other Marines on the gun team. Within 10 seconds, we were ready to fire.

The field telephone at the bunker rang. I answered and heard the voice of the Commanding Officer of Delta Company, telling me that he was contacting battalion headquarters and requesting permission to fire. I told him that we were ready. It was essential to ensure that there were no friendly forces who might be harmed by my fire. Almost immediately, the CO phoned back to tell me that I was cleared to fire.

The rocket site was about 3,300 meters away, far outside the specified maximum effective range of the 106. The .50 caliber spotting tracer did not have the range to reach the rocket site. I pointed the 106 at the rocket site, raised its elevation, and pulled the firing knob to fire the spotting rifle which was mounted on the top of the 106 and bore-sighted to be in perfect alignment. The tracer did not reach mid-arch before it burned out.

Rockets were being launched at a furious pace. The enemy gunners knew that they had only a short time to release their rockets and run before the wrath of Americans with big guns fell hard on them. Counter-battery fire from artillery would take too long to arrange and adjust. In the several minutes that it would take to get an artillery round in the air and headed toward a rocket site, all the rockets would have been fired and the enemy gunners gone. Seconds counted. Each rocket fired could cost lives and valuable aviation assets on the ground in Da Nang, 15 miles away.

I had never fired the 106 against targets as far as this rocket site. The spotting rifle was of no help. I asked the other Marines to keep an eye out for the impact and fired a high explosive round from the 106 toward the rocket flashes. The yellow fireball resulting from my round was 1,000 meters short and to the left of the rocket site. I adjusted fire with the two wheels that controlled the azimuth and elevation of the rifle and fired again. A patrol in Happy Valley could see the rocket site and the impact of my fire. The patrol radioed an adjustment. My next round was 200 meters to the right and 100 meters long. I made small adjustments and fired my third round.

The third round hit the rocket site directly, causing a fireball explosion and sending rockets careening off the site into uncontrolled flight. One of them lifted off, then turned back to the rocket site in a skyward loop, ultimately landing on its own launch team and exploding in a ball of fire, causing a secondary explosion.

I could hear the radio conversation between the command center on Hill 41 and the patrol adjusting fire: "Is the 106 on target?" and the excited response that violated radio protocol, "The 106 is kicking the shit out of 'em." I fired seven more rounds into the rocket site. There were reported to be a total of five secondary explosions where things on the ground at the rocket site exploded as a result of the explosion of my 106 rounds. The tower on Hill 41 and the patrol confirmed the secondary explosions and accuracy of the 106 fire.

By the nature of counter-battery fire, the enemy gets off some rockets before the counter fire can begin. On this night, 28 122mm rockets fell on the Da Nang airfield, causing the destruction of one aircraft and damage to six others. No personnel were reported injured in the attack. The enemy gunners were not so lucky. The enemy had hoped that their rockets would land among Americans in Da Nang, killing many with their high explosive charges by propelling slivers of metal shrapnel through flesh and bone. The irony is that the enemy rocketeers suffered that same fate when the rockets exploded at their launch site in Happy Valley rather than on the Da Nang airfield. In addition to their own rocket detonations, high-explosive 106 mm rounds sent concussion and shrapnel through the rocket site, causing the same damage that they sought to inflict on Americans in Da Nang.

The Move Toward Da Nang

Around 0200 hrs. on the morning of February 6, Combined Action Platoon (CAP) Bravo-3, in the hamlet of Duong Lam (1) near the Tuy Loan River, received incoming mortar and small arms fire, followed by a ground attack. Nearby, CAP Bravo-4 was attacked by two companies of NVA. The CAP units called for help.

A squad from 1/7, along with two tanks in support, responded to the call from CAP Bravo-3. The reactionary force came under heavy fire as it approached the besieged CAP unit. Both tanks were disabled by rocket-propelled grenades. A short time later, two more tanks arrived along with South Vietnamese soldiers of the Army of the Republic of Vietnam (ARVN) to join the rifle squad from 1/7 in the rescue of CAP Bravo-3.

The reactionary force pushed into the village and fought its way to the compound of CAP Bravo-3. Marines from the reactionary force found the bodies of two NVA soldiers and captured three more. The prisoners claimed to be from the 3rd Battalion, 31st NVA Regiment and stated that their ultimate objective was Da Nang.

At about the same time, nearby CAP Bravo-4 was attacked by two companies of NVA. Marines from Delta Company 1/7, based on Hill 41, were on night patrol in the area and came to the defense of Bravo-4.

A short time later, CAP Bravo-1 came under attack. A reactionary force from Alpha Company, 1/7 based on Hill 10 and conducting night patrols, moved to assist Bravo-1.

The NVA were unsuccessful in defeating any of the CAP units and suffered 34 killed at Bravo 3 and 4, and an additional 18 killed at Bravo-1. The enemy force would now turn toward its ultimate objective, the Da Nang airbase about 15 miles to the North. Standing between the NVA and Da Nang was Hill 41, a small firebase with only one Marine rifle company. They hoped to slip past Hill 41 undetected but were prepared to attack and destroy the firebase if necessary.

Local Viet Cong informers had provided information about the patrol patterns of the Marines. The patrols left their firebase before dark and set up a circular defensive position, where they ate their dinner from green cans and waited until dark. The patrols used the cover of darkness to move to a place where they would conceal themselves and wait to ambush VC who might come along. In the morning, the Marines would leave their hiding places around 0600 and return to their base. They always arrived back at their base for breakfast before the chow hall closed at 0800.

To avoid the Marine patrols, the NVA waited until 0800 to begin their movement toward Da Nang. As the enemy force crossed the road leading from Hill 41 to Da Nang, they did so in small numbers at a time, to avoid being observed. As they were doing so, five Marines, three with rifles and two with technical equipment appeared. The three Marines with rifles fired upon the five or six NVA who were caught in the open while crossing the road.

There would be no slipping past Hill 41.

HAPPY NEW YEAR

After sunrise on February 6, I left the sandbagged bunker on Hill 41 and walked the short distance to the tent that I called home. On top of the bunker was a 106 recoilless rifle. Its field of fire included the open rice fields, the village at the base of the hill, and Charlie Ridge, a mountain range west of Hill 41. I spent my nights in that bunker, ready to fire the 106 and keeping that section of the hill's perimeter free of Viet Cong intruders. I was exceedingly lucky to have such good duty. My military occupational specialty code was 0351, assault man. The basic rifleman was assigned the MOS code 0311. As an 0351, I had been trained in various infantry weapons, including the 3.5-inch rocket launcher which some people might refer to as a "bazooka," flamethrowers, demolitions and the 106 recoilless rifle. The 106 was a crew-served weapon weighing 462 pounds. Although we did have a mobile gun that we took into the field on a tracked vehicle, most of the 106's were in fixed positions at combat bases. That prevented me from having to patrol with the rifle squads and spend my nights walking several miles to set up ambushes for Viet Cong attempting to sneak into our area. It was also a lot safer than the life of a rifle squad "grunt." It was not uncommon for an 0351 assaultman to report to his assigned unit and be assigned to a rifle squad with the explanation that they had all the 0351's they needed.

From time to time I would join a daytime patrol from Hill 41 just as a change of pace. It made me appreciate the relative comfort of the 106 bunker.

For the riflemen of Delta Company, it had been an unusual night. The squads of eight to ten men had left the hill before nightfall and moved off in separate directions toward their patrol areas. Before sunset, they stopped to set up a 360-degree defensive perimeter, prepared their dinner of C rations, and waited for nightfall. After dark, they had used the darkness to conceal their movements into their ambush positions along the trails and rice paddies of the approaches to Da Nang and the areas from which the enemy could attack the major military complex at Da Nang with their rockets.

About 0200, most of the squads had been instructed to leave their ambush positions and move to support a Combined Action Platoon (CAP) which had come under enemy attack. The CAP units were small, with only a squad-size complement of Marines. The Vietnamese Popular Forces troops assigned there were mostly useless and would hide when the shooting started, leaving the defense of their small base to the Marines. The Marines in the CAP units depended upon support from rifle companies in the area when attacked.

There were six CAP units within the operating area of the 1st Battalion, 7th Marines. During the 13 months I was assigned there, all of them had been attacked and many of them overrun. It seemed to me that serving in a CAP unit was a form of Russian roulette combined with musical chairs. At the time, I had no idea that one day I would be assigned to a CAP unit.

Four of the six CAP units in the area were attacked during this night. CAP Bravo-4, nearest to Hill 41, was assaulted by two companies of NVA. Marines from Delta Company moved to assist CAP Bravo-4 and repelled the assault, killing 34 NVA. Alpha Company on nearby Hill 10 sent a reactionary force, supported by tanks, to help defend CAP Bravo-3.

The attacks on the CAP units drew the resources of Delta and Bravo Companies from their mission of preventing movement of enemy forces through the area. Although the intervention of the reaction forces prevented the CAP units from being overwhelmed, it opened the door for the enemy to move unseen and unmolested toward Da Nang.

After dawn, the exhausted riflemen of Delta Company began to return to Hill 41. They arrived, one squad at a time, at the personnel gate, where they negotiated the winding path through the barbed concertina wire that protected the perimeter. They approached the hill, just as they had left, with 15-meter intervals between each Marine. As they entered the perimeter, they removed the magazines from their M16 rifles and pulled back their bolts, ejecting the round in the chamber and catching it in their hands. They hurried to the mess tent, where breakfast was being served. After being awake all night, patrolling for miles, and participating in the fight at the CAP unit, they were looking forward to a breakfast of bacon, eggs, and coffee followed by a couple of hours of rest.

The Marines in the mess tent early on this Tuesday morning included some who had enjoyed the rotating duty of remaining at the combat base the night before, as well as those coming in from night ambush patrols.

Sergeant Glenn Lunsford had been wounded three days earlier by grenade shrapnel and had not been required to participate in the patrols of the previous night. He was having breakfast with several other Marines when Lieutenant Henry Wright hurried into the tent and ordered several Marines, including Sgt. Lunsford, to get their gear and prepare to move out as a reactionary force. The Lieutenant directed one of the cooks to go to the tent of the 106 recoilless rifle crew and tell them to man their gun.

After sunrise, I left the sandbagged bunker on Hill 41 and walked the short distance to the tent that I called home. I stood by the canvas cot in the tent that I shared with the rest of the 106 team assigned to Delta Company, 1st Battalion, 7th Marines. I had just removed my helmet and flak jacket and was looking forward to some rest when one of the cooks from the mess hall next door entered the tent to relay a message. He spoke to me, saying : "They want you back on the gun."

Thinking it odd that I would be sent back to man the recoilless rifle during the day, I left my flak jacket and helmet on my bunk, put on my soft cover cap and headed back to the bunker. Almost all combat occurred at night. The Viet Cong were at a significant disadvantage during the day and preferred to come out at night. During the day, the local Viet Cong pretended to be innocent farmers. When the sun came up, "Mr. Charlie" resumed his peaceful peasant role, perhaps even going to his job working for the Americans at one of the military installations.

As I walked to the bunker, I saw a tank begin to move from the place it had spent the night on the hill toward the front gate. By the time I had gotten to the bunker, the tank was moving down the road that led away from the hill toward Da Nang. It was firing its 90mm gun over a tree line and into rice fields about a thousand meters away. I picked up the field telephone and spoke to the command bunker, informing them that the 106 recoilless rifle was manned and ready. The voice from the command center instructed me to fire where the tank was firing.

Ed Sikes joined me on the gun and assisted as I watched the tank fire its next round. I felt naked and a bit vulnerable without my helmet or flak jacket. I had no idea that I was about to get into a serious firefight when I left my tent to return to the gun and was surprised to see this development. I was even more surprised when the tank suddenly stopped moving forward, and smoke began to drift out the top hatch. It had been hit by a rocket-propelled grenade (RPG). When a crewman climbed out the hatch, he was shot by a sniper. Other members of the crew had been wounded but escaped through a hatch in the bottom of the tank.

Pagoda

I looked around to see if I could determine where the RPG that took out the tank came from. I knew that it had to be relatively close to the tank and not as far away as the tree line, which was several hundred meters away in the direction the tank had been firing. Near the intersection where the road from Hill 41 met the road to Da Nang stood a Buddhist pagoda. We respected the religious shrines of the locals and made it a point not to damage them needlessly. There was now a need. The most likely cover and concealment from which the enemy could have fired the RPG was the pagoda. My concern for the cultural sensitivities of the locals lasted about a nanosecond as I lined up the pagoda in the optical sight of the 106 and hit the fire button. Marines were being killed, and cultural niceties were out the window.

Reactionary

Behind me, Marines were running from their tents and bunkers carrying their rifles and hurriedly putting on helmets, flak jackets, and pistol belts. They quickly loaded onto two trucks that were waiting near the gate with their engines running. A reactionary force was being organized to rush to the aid of Marines in trouble.

Every morning at first light, before vehicles were allowed on the road, a team of two engineers, with a three-man fire team for security, swept the road for mines. On this morning, the team observed several North Vietnamese Army soldiers run across the road ahead of them in a forested area. The three-man fire team opened fire on the small enemy force. What they did not know was that the NVA they engaged were part of a force of approximately 1,500 enemy soldiers attempting to slip past Hill 41 to attack the large US airbase in Da Nang. The mission of 1/7 was to prevent exactly that.

When the fire team realized that the enemy force was far larger than they had expected, they radioed the command center on Hill 41 to report that they were in contact with a large number of NVA and needed help. Most of our contact with enemy forces in the area of our patrol base was with local Viet Cong. The presence of NVA troops, particularly during daylight, was unusual and an indication that something very different was happening. What that was had not yet become apparent to either the mine sweep team or the command post on Hill 41. What was evident however was that the five Marines on the mine sweep detail needed help, right then.

A reactionary force often included everyone on the hill not engaged in other duties directly connected to the fight. The 30 or so Marines of the reactionary force locked and loaded as the truck headed out the gate past the disabled tank. The truck reached the T intersection at the bottom of the hill where the road from the hill intersected the road running past the hill toward Da Nang. The Marines in the trucks piled out and formed a line abreast facing the tree line where the enemy was thought to be.

It was becoming clear to me that something serious was in the works, but I did not know and did not ask, because I knew that the command center did not have time to explain it to me. Although I did not know what I was shooting at, I followed my orders and fired over the tree line into the rice fields beyond. My rounds impacted and exploded out of my view behind the tree line.

A few days before, my team had decided to reconfigure our bunker. We removed the sandbags that were stacked three or four rows high around the top of the bunker to protect against small arms fire and shrapnel. We intended to replace that protective screen of sandbags with a new design but had not yet done so. The 106 sat on its tripod on top of the bunker, nakedly exposed. The unfortunate timing of that decision became apparent when the first small arms rounds began to impact the bunker. A bullet passing close by causes a sound like that of a bullwhip. It is the sound of the bullet breaking the sound barrier as it goes past.

At first, the sound of the firefight was limited to the area where the reactionary force was moving across the rice paddies toward the tree line. Then the CRACK of rifle fire directed toward my position began. It came from the general direction of the firefight in the rice paddies, but closer. The NVA in the tree line were at least a thousand meters from my bunker, far outside effective rifle range, except perhaps for snipers. There were two villages in that direction, one near the place where the reactionary force had entered the rice paddies and another closer to Hill 41. The fire directed at my position seemed to be coming from both those directions. The incoming small arms fire was light at first but gradually grew in volume and increased in the number of directions from which it came.

Sikes and I took cover behind the only thing on the top of the bunker that we could shelter behind in the absence of the parapet that we had removed a day or two before and had not yet replaced. Ammunition for the 106 was stacked in cardboard tubes, one round per tube, against the legs of the tripod upon which the gun was mounted. I continued to direct my fire to the area where I had first been instructed to fire although I could not see the impact of my rounds. My target was obviously NVA troops in the area behind the tree line. I knew that there was a small South Vietnamese Army unit in that area and worried that my fire might be too close to their base. I need not have worried. They had run away at the approach of the NVA and done nothing to alert us of the presence of the enemy.

To aim the 106, it is necessary to look into an optical sight on the side of the rifle. In order to look through that sight, it was necessary to sit up or kneel alongside the gun to get your eye level with the sight. Both Sikes and I were trying to get as small as possible by lying on our stomachs and taking cover behind the short stack of ammunition leaning against the tripod. After I had fired a round, Sikes would reload the rifle. To do that, he had to open the breach, grab the spent shell casing, pull it out of the gun and throw it behind the bunker, pick up a new round, load it into the gun, and close the breach. That reloading process took a few seconds, during which he had to stand. When I raised myself up from behind cover to aim, and notably when Sikes rose to reload, the volume of fire increased. At least one enemy sniper was waiting for those moments. Probably the one who killed the tank crewman. Sikes and I took turns presenting ourselves as a target. I would rise to aim then drop down after the round was fired, then Sikes would stand to reload. I did not have to aim with the optical sight for each shot. There is a second way to determine where the 106 round will impact.

Mounted on top of the 106 recoilless rifle is a .50 spotting rifle that fires a tracer round with a white phosphorous tip. The red circular firing button mounted on the side of the rifle can be pulled to fire the .50 or pushed to shoot the 106. The spotting round shows the gunner where the 106 round will strike by releasing a white puff of smoke at the impact point. When the gunner is satisfied with the aim point, he can fire the 106 round which will impact in the same position. While that system worked well up to the designed effective range of 1,100 meters, it was ineffective for ranges greater than that because the tracer burns out before it reaches the target.

The Rice Field

The Reactionary force was moving across the rice paddies toward the tree line about 700 meters in front of them. The officer in charge of the reactionary force, Second Lieutenant Henry Arthur Wright, had recently graduated from the Naval Academy in Annapolis and had been in Vietnam three weeks. He deployed the 30-man reactionary force in a line abreast formation and intended to advance toward the enemy using the infantry tactic of fire team rushes. The main body of the force would lay down a base of fire while 3 man fire teams took turns rushing forward for a few yards then hitting the deck and laying down fire while another team advanced. Each time a fire team stood up to rush, it was cut down by withering enemy fire. The reactionary force knew that they were going forward to engage an enemy force of undetermined size. The extreme volume of fire coming from the tree line was the first indication that the size of the enemy unit was enormous. In fact, the Marines were attacking an enemy force 50 times its size. The reactionary force was taking casualties at a rapid and unsustainable rate. If they continued the assault on the tree line, none of them would remain alive long enough to reach it.

Back on the Bunker

It seemed that the enemy was moving forces forward to confront the Marines assaulting toward the tree line and suppress supporting fire from Hill 41. The only weapon on Hill 41 with the range to support the Marine reactionary force was the 106 recoilless rifle, with which I continued to fire high-explosive rounds into the enemy force behind the tree line. A universal combat tactic is to concentrate your fire on the enemy weapon that is causing you the most damage. In this case, that would be the recoilless rifle. At first, the fire was coming from one direction, that of the villages between Hill 41 and the tree line. Then small arms fire came from other directions, including other villages. With each new enemy position established, the volume of fire arriving at the 106 bunker increased. Bullets were striking the ammunition that I was taking cover behind. The sandbags that made up the bunker were absorbing small arm rounds causing sand to dribble out of them in several places. The CRACK of bullets passing close by was increasing in frequency

There was one enemy sniper that was particularly worrisome. It took Sikes about seven seconds to reload after each shot fired. As he finished reloading and jumped back down to take cover, a CRACK would be heard very close by. On some occasions, a tracer would pass just where his head and upper body had been only two seconds before. Each time one of us exposed ourselves to fire or reload, we drew fire. That fire was close but not precise, except for one sniper, who was very precise. Fortunately, he or she did not have quite enough time for a carefully aimed shot and smooth trigger pull during the time it took Sikes to reload. The sniper needed a couple of seconds longer. On one occasion, Sikes had trouble extracting the spent shell casing, causing a delay of a few seconds. Sikes knew just how much time he had to reload before the enemy bullet arrived. Sikes realized that the time it was taking to free the stuck casing would cause him to overstay his standing position. When he got to the point where he knew the sniper round would be coming, he abandoned the reload attempt and dived for cover. Just as he did a loud CRACK was heard as the snipers bullet arrived, just missing him.

Ammo Resupply

In addition to the few rounds that we kept on top of the bunker alongside the gun and ready for immediate use, several other rounds were stored inside the bunker. A fire mission against rocket sites usually required a few high-explosive rounds. This day would require much more, which meant that we would need more ammo brought from the ammo storage bunker. It did not take long to burn through the dozen or so rounds on the gun bunker. We called the command post and informed them that we needed an ammo resupply.

Our bunker stood in an open area of the hill. To get to it, you would have to cross a good deal of open ground. Sikes and I had cover on the bunker except when we had to expose ourselves to fire or reload. There would be no cover for anyone bringing ammo to the gun bunker. By this time, we were taking fire from several directions. All that fire was concentrated on the only weapon on the hill that was firing at the enemy forces, the 106 recoilless rifle.

In the days of sailing ships, young boys were assigned the task of resupplying the cannons with gunpowder. They would run between the gun positions and the powder magazine. They were known as "powder monkeys." Opposing forces did their very best to shoot the powder monkeys. If you could prevent the resupply of ammunition, you could render the cannons useless against you. Some things never change. Our powder monkey was a Marine nicknamed “Wisenheimer.” I don't know how Jerry Mensinger from Pennsylvania got the name “Wisenheimer,” but it was not meant to be derogatory. Wisenheimer was well-liked and could always be relied upon to be there when he was needed.

Not long after I called to request a resupply of ammo, Wisenheimer came running across the open area toward the bunker with a 106 round on each shoulder. Immediately the enemy small arms fire increased, and bullets could be seen kicking up sand around Wisenheimer as he sprinted like a running back with the goal line within reach. A 106 high explosive round weighed more than 20 pounds. When he reached the gun bunker, he lurched his shoulders forward propelling both of the rounds onto the sandbagged roof and said, "I'll be right back with some more." Instantly, he was off again, running a zig-zag path up the hill with enemy bullets kicking up red dust around him. Wisenheimer did come back several times. Each time, he brought with him two rounds of ammunition and each time he had to run the gauntlet of enemy fire.

Even with our powder monkey delivering ammo to the bunker, we faced an ammo shortage. I was burning through all the ammunition on the hill. We would need more, and we would need it soon. That information was relayed to the command bunker, who sent the request to division headquarters in Da Nang. Marines in Da Nang rushed to the ammunition storage facility there, broke out a pallet of 106 ammunition, and took it to the airfield where the Marine Air Wing had arranged for a helicopter to be waiting for it.

Back at the rice field

Things were not looking good for the reactionary force. Their initial mission was to come to the aid of the mine sweep team and to engage the enemy force they found there. Now, they found themselves as a blocking force, preventing the North Vietnamese Army from breaking through to attack the airfield at Da Nang. They were the only thing standing in the way of the enemy. Not only were they outnumbered 50 to 1, but they were also in a poor tactical situation. The enemy was firing from a position of cover and concealment in the tree line. The Marines were in open rice paddies. The new second Lieutenant found himself in command in a battle for which the Naval Academy and his limited experience had not prepared him. Fire team rushes were not working, and casualties were mounting.

Sergeant Glen Lunsford was near the end of his second tour in Vietnam. Even one tour in a Marine line company was tough on a person. The chances of becoming a casualty, that is, being killed or wounded, was about 85 percent in just one tour. As losses mounted, Lieutenant Wright became one of them, killed by enemy rifle fire. That left Sgt. Lunsford in charge. Lunsford made the wise decision to alter the tactical situation by withdrawing from the open paddies and calling in artillery and air strikes. As the Marines were moving out of the paddies, Lunsford was also shot and killed.

I was frustrated by firing over the tree line when the action was in the paddies, and the enemy force inflicting casualties was in the tree line, not behind it. I called the command bunker and asked if I could fire into the tree line. The answer was "no, keep firing where you are firing." I did as I was ordered although I thought I could have been more effective otherwise. There was another reason for my caution. The tracer of the .50 caliber spotting round on the 106 burned out before it got as far out as the tree line. I would have to adjust fire by shooting a 106 round and watching where it landed. If my round fell short, it could land on the Marines in the rice field. I asked the command center if the reactionary force could adjust fire for me. The command center advised me that the Marines in the rice field could not see the impact of my fire.

Back at the Bunker - ammo resupply

I had fired almost all the 106 ammunition on the hill when the sound of helicopter blades was heard. A twin-rotor Marine Sea Knight helicopter was flying toward the hill with a pallet of ammo slung underneath it. The small arms fire directed at the 106 bunker shifted to the helicopter as it approached. Ignoring the incoming fire, the helicopter hovered over an open area on the hill and descended until the pallet rested on the ground. The pallet was released and the helicopter flew away, taking fire as it did. It did not get far. Just after the helicopter lifted, it lost power. The Sea Knight had been damaged by enemy fire and had to land along the road near the disabled tank. The crew of the helicopter jumped out and ran toward the hill yelling, "open the gate!" The crew was able to reach the gate safely and found shelter on the hill but did not seem very happy about being temporary residents of Hill 41.

Artillery and Air Support

When the Marines in the reactionary force were in contact with the NVA, they were too close to the enemy to use artillery. When the reactionary force withdrew from the open area, that changed. Sgt. Lunsford's last order opened the distance between the Marines and the enemy enough to utilize artillery and air support. Although the NVA were inflicting significant casualties on the Marines, the enemy did not do so without penalty. The Marine force was inflicting casualties on the enemy within the tree line, and my 106 was pounding the NVA behind the tree line. Any hope that the enemy had of breaking through the Marine defense and attacking Da Nang came to an end when the F4 Phantoms arrived.

Earlier in the morning, things were looking encouraging for the NVA. The South Vietnamese Army troops fled their compound without raising the alarm as the NVA approached, leaving the path to the vital American target open to the NVA. Nothing blocked their way, except a three-man Marine fire team and two engineers armed with a metal detector.

At that point, a tank from Hill 41 started firing into their advancing force as it used a tree line between the ARVN base and Hill 41 to cover their movement. The tank was quickly dispatched by an enemy soldier with an RPG, pre-positioned to cover the only road from Hill 41. Then the 106 recoilless rifle on Hill 41 began to pound their advancing column with high-explosive rounds. Unlike the tank, the 106 proved difficult to eliminate even as their supporting riflemen and snipers poured fire toward it.

When 30 Marines came from Hill 41 and attacked across the rice paddies, the NVA fired from the tree line, causing massive casualties to the ridiculously small force sent against it. The Marines of Delta Company on Hill 41 were doing precisely what they were there to do. They were blocking an enemy force from reaching the strategically important and massive airbase in Da Nang and defending their own combat base in the process.

Despite their far superior numbers, the NVA knew that their mission was in jeopardy. The critical element of surprise had been lost. They had been detected while still 15 miles from their target.

After the survivors of the Marine reactionary force pulled back from the rice field, taking their wounded with them, the whistling death sound of artillery arrived. Exploding 105mm and 155mm howitzer shells joined the 106 shells raining down on the NVA. The advance of the enemy column was clearly stopped, and any hope of completing their mission was over. Now they hoped to escape with their lives. Few of them would be able to manage that.

When the artillery stopped, the sound of jet aircraft announced an even greater danger. Soon, F4 Phantom jets were diving on the enemy column. They came from the airbase the NVA had hoped to destroy. They brought with them high-explosive bombs and napalm. The NVA were now in headlong flight back in the direction from which they had come. Soon, they were out of range of the 106 on Hill 41, but not outside the range of artillery and air strikes. The Phantoms continued to pursue the enemy column, leaving a trail of dead bodies as they fled.

Coordinated blocking force

At the headquarters of the First Marine Division in Da Nang, the staff was following the action closely and trying to stay ahead of it in planning. The NVA force had been stopped, then turned, and were fleeing with air strikes punishing them along the way. Turning them away was not enough. The primary goal was to protect Da Nang. The end game was killing as many of the enemy as possible.

In the early afternoon, the attacking jets could be seen diving on the enemy column like a flock of seabirds diving on a moving school of fish. The bad juju was just beginning for the fleeing NVA soldiers. At Marble Mountain near Da Nang, Marines were loading up on amphibious tracked vehicles, known as "AMTRACS," to move into the path of the fleeing enemy and act as a blocking force. The NVA who survived the firefight near Hill 41, the 106, the artillery, the air strikes, and who had been fighting and running all day, would soon find a large group of determined Marines blocking their escape path.

Larger Enemy Effort

Beginning on Jan 30, attacks had been launched against over 100 locations in South Vietnam. This was an all-out effort by the North Vietnamese and Viet Cong. Across the country in over 100 cities, towns and military bases were being attacked. The Viet Cong who had been laying low now came out to fight. The enemy fire directed at Hill 41 was not from the NVA main force; it was from the villages around the hill. It is unknown whether the shooters were local Viet Cong or NVA who had been welcomed into those villages to attack the Marine combat base. Viet Cong were usually not very good marksmen and did not have the training that soldiers of the North Vietnamese Army had.

After the Battle

It was afternoon when things calmed down and Sikes and I were able to take a look at the damage to our bunker. Sand had dribbled out in several places where bullets had ripped holes in the sandbags. On the cardboard casing of the 106 round I had used for cover, there were three gouge marks right next to each other, all within a span of three inches. Bullets had ricocheted off the edge of that round. Those bullets passed within a couple of inches of my head!

I was aware of the intensity of the incoming fire and how close some of it was. All Marines know the CRACK sound of small arms fire passing nearby. In boot camp, everyone takes a turn at "pulling butts." The targets on the rifle range are mounted on frames that are raised and lowered from below a dirt berm. Recruits raise the frame and target above the berm that shields the "butts" from the rifle fire. The "butts" is the area into which the target is lowered to have the impact of the bullet recorded and a patch applied to the bullet hole before it is raised again to receive the next round. The bullets pass about four feet above the recruits in the "butts." Whether by design or not, that process ensures that every Marine knows the sound of a bullet that passed close to him. Sometimes you can see tracers pass by you. That is a pretty frightening experience.

Left for Dead

When the reactionary force withdrew from the rice field, they did so under withering small arms fire from the tree line. They brought their wounded with them but were not able to bring out all of their dead. The bodies of several dead Marines remained in the rice field after the fight and through the night. The enemy had been routed, and the main body fled, but some were still around and continued to pose a threat that prevented the Marines from operating outside the wire of their combat base on Hill 41. The following morning, Marines returned to the rice field to retrieve the dead.

One Marine had suffered a head wound and appeared to be dead. When the order to withdraw from the rice field was given, his comrades saw that he was shot in the head and saw no signs of life. They left his body there. When the Marines found him the next morning, they were astonished to see that he was, in fact, still alive! In addition to his head wound, he had been struck by fragments of friendly artillery called in against the nearby enemy force. The shocked Marines radioed for a Corpsman, who ran down the road from Hill 41 with his medical bag and hijacked the bicycle of a passing local to get to the rice field faster.

The wounded Marine had been surrounded by enemy soldiers who had searched the bodies of the dead Marines and removed weapons, ammunition and equipment. They had taken his canteen and other equipment from his belt but did not realize that he was alive. A medevac helicopter arrived to take him to Da Nang for medical care. Months later, we would receive a letter from his father informing us that he was still alive and recovering from his wounds in a hospital back in the States.

AMBUSH AT THE WATERFALL

My grandfather, who served in the army infantry in World War I, once told me that when the leadership wanted to know whether the enemy was around, they would send a few troops "over the top" of the hill. If the enemy was nearby, the soldiers would be shot. Even as a child, I thought that was a horrible thing to do. Surely, he had exaggerated. The army would not sacrifice the lives of its troops just to use them as the infantry equivalent of the coal mine canary.

I remembered that story as I moved through the high elephant grass on a hillside near the special forces camp at Thuong Duc. I was "walking flank." My squad was taking our turn moving parallel to the main body of our battalion-size operation, "Mameluke Thrust." It was late May, and the heat was already oppressive even if you were standing in the open. The elephant grass absorbed and retained the heat like an oven. There was no way that you could move silently through it and keep pace with the main body, which was moving along a river bed below. Moving up and down the hills along the river bank added a good deal of distance through terrain much slower to transit than the level and open river bed ... unless the main body happened to be in a portion of the river that required wading.

The tall grass, overhead high, had sharp edges that could leave tiny but painful cuts on exposed skin. It was impossible to see more than a couple of feet because the grass was so thick. It was not a problem to know where the other squad members were because the sound of the rustling grass clearly gave away their positions. The enemy could be waiting for us to approach, listening to us move toward them, and we would not know until they opened fire at close range from concealed positions. I now realized that my grandfather's story was not an exaggeration. Things had not changed since 1916, and the army was not the only branch of the military that practiced "reconnaissance by being shot."

The concept of the practice was sound, although primitive. The flank squad moved just ahead of the main body. It was almost inevitable that at some point, the enemy would be lying in wait to ambush the main body. The enemy could choose the place of the ambush and take advantage of the terrain to provide cover, concealment, and other tactical advantages. A perfect ambush allows the ambusher to trap the ambushees in an area where they cannot escape or easily mount a counter-attack. The specific place where the ambushed party would be most vulnerable and where they could expect the enemy to open fire is known as the "kill zone." It is a place that you do not want to be.

Marines are trained to charge directly into an ambush. The natural reaction would be to try to find cover and escape. However, in a properly planned and executed ambush, it is likely that there is no adequate cover and no route of escape. The enemy chose to open fire in that place for a reason. You are in their kill zone. If you try to escape it, you will probably be shot in the back by an enemy soldier who feels no threat from you and can take his time to aim carefully as you feebly and unsuccessfully attempt to run away. It unnerves the enemy when you charge directly into them.

If you knock down a wasp nest and you know that the wasps will all fly away in terror, you have no reason to fear being stung and can take your time. If you know that the wasps will swarm and attack, your attitude will be quite different. Even if you kill most of them with your can of wasp spray, a few survivors will attack viciously, and you are likely to be stung. When the enemy can catch Marines in a well-planned and executed ambush, it is not like calmly shooting the ducks moving across the pond in the carnival tent. When an enemy opens fire on a Marine, he has just started a confrontation that will cause an immediate and violent reaction. The enemy, in a superior tactical position, will probably prevail. However, the survivors of the initial attack will run directly toward him and will be the meanest human beings the enemy has ever seen and will be intent on killing him. The Marines who survive the initial ambush fire and charge into the enemy force may also be killed, but they will inflict as much damage on the enemy as possible before they are hit by enemy fire.

Every enemy soldier should wonder if he will be part of that damage. An enemy soldier concealed in a fortified position with a tactical advantage should feel that he is about to start an action that just might result in his death. That should make his heart race and his trigger finger tremble. That makes him less accurate with his fire and more likely to run away from the counter assault. Shooting at a Marine has consequences akin to hitting a rattlesnake with a stick.

The hills were not all elephant grass; there were open areas as well. In the open, it was oppressively hot, but not as hot as the oven that was the elephant grass. We were limited to three canteens of water a day. We wore flak jackets that would protect us from shrapnel produced by exploding grenades and mortars, but not from gunfire. Our helmets provided the same sort of protection. The added weight of both made it more difficult to move. The weight of an infantry kit, ammunition, grenades, three canteens, first-aid kit, bayonet, and of course a pack did not make it easier to move up and down the hills.

Heat exhaustion was a constant and common problem. I had learned early in my time in the Marine Corps that you are physically capable of more than you think. I also learned to never quit and to push myself to my physical limits. Sometimes, in the bush, you reached those limits. We were in an open area moving forward in a line abreast tactical formation. I was baking in the sun. I became light-headed and knew that I was near the end of my physical capability. On the edge of collapsing, I pushed on. Just then the Marine to my left collapsed from heat exhaustion.

We halted while we attended to our heat casualty. We had no corpsman with our squad. We did have a Lieutenant with us, unusual for a squad size patrol, who gave some of his precious water to the Marine who had collapsed. I remember that the water in the Lieutenant's canteen was Kool-Aid. It was common to add Kool-Aid to our water, particularly if the water had been treated with halazone to kill harmful bacteria. The Kool-Aid masked the awful taste of the halazone. All of us were suffering from heat exhaustion. The Marine who collapsed was not much worse off than the rest of us; he was just the first to succumb.

Being able to push yourself past normal physical limits had the advantage of allowing you to accomplish more. It has the disadvantage of allowing you to damage your body. There is a saying in the Marine Corps today that "pain is weakness leaving your body." Pain and discomfort are communication from your body that you are endangering it to the point that it may become damaged. We were at that point. We were spent. Unable to think clearly, barely able to put one foot in front of the other, moving through steep terrain with the need to keep up the pace to stay ahead of the main body, and needing to stay alert to people hiding in the bush waiting to kill us, we were unable to fulfill our mission requirements.

Another squad was brought up to relieve us. We rejoined the main body sweeping along the banks of the Song Vu Gia River. After flank duty, walking along the river, sometimes wading through it and sometimes moving through the boulders of the dry creek bed, was almost pleasant.

There were rifle companies of Marines and the command unit. I was assigned to the command unit, which was usually in the middle of the group so that the battalion commander, a Lt. Colonel, could best exercise operational command. My squad was used to fill in where needed and assigned ad hoc missions such as manning listening posts, walking flank patrol, and responding to situations where manpower was immediately required to assist one of the rifle companies.

Moving along the river bed, we came to a bend in the river overlooked by a steep hill. Sounds of a firefight came from the area of the hill. The flank squad had performed its function. The Marines around me and I took cover behind the boulders in the river bed as we began to receive fire from the hill. The flank squad had discovered an enemy force dug in on the hillside waiting for our main body to enter the kill zone of their ambush site. The kill zone would have been in the rocks of the river bed just under the hill, where the enemy could have fired down onto the Marines from concealed and fortified positions. Directly below the entrenched enemy positions, the Marines would have no adequate cover and would be trapped by the steep sides of the river bed. It would have been a textbook ambush and would have been devastating. The enemy was denied the ability to inflict maximum damage to us because they were discovered and engaged by the flank squad before the main body reached the kill zone.

The flank squad had suffered casualties. Several men from my platoon were drafted to move up the hill and rescue the remainder of the flank squad and bring their casualties back to a safe area where they could be treated for wounds and protected until they could be medically evacuated.

The sound of a firefight did not come only from the flank squad. Gunfire could be heard from some point ahead of my position. The enemy must have been disappointed that the main body of our force had avoided their kill zone. Even so, we took some casualties as the NVA fired down on us from the hillside as we took cover among the rocks. I was further back in the column and in good cover. The problem for the enemy, especially during the day, is that they need to strike hard and move away from the area fast. Our infantry can usually call in supporting fire from artillery and aircraft. However, our position was outside the range of our supporting artillery. Air support was hard to get probably because it was in short supply. We called for air support and hoped it would be available.

In the meantime, several members of the flank squad had been wounded. We needed to get them off the battlefield and to Da Nang, where they could receive medical attention. An Air Force helicopter flew into the landing zone but departed without touching down because of the massive volume of fire directed toward it. The pilot called down and said that the landing zone was too hot for him to land. He asked how badly wounded our casualties were. When he was told that some of them were critically wounded and needed immediate transport to a hospital, he decided to come back and pick them up. I was surprised that an Air Force helicopter was performing our medevac; I had never seen that before. We always had Marine helicopters fly those missions.

When the twin rotor Air Force helicopter returned, the fire

; just as intense, maybe even more so. He landed the helipter on a flat spot on the rocky river bed and remained there while Marines ran across open ground to the aircraft, carrying the wounded, and hurriedly loaded them aboard. Holes could be seen appearing in the skin of the helicopter as it sat with its rotors turning. It took only a few seconds to load the casualties. The helicopter immediately lifted and flew away with the NVA shooting at it and the Marines firing at the NVA to try to keep their heads down while the medevac flew out of range.

Not long after that, our air support arrived: two silver-colored F4 Phantom jets with Air Force markings. They made several passes at the enemy position on the hill. The Air Force jets seemed very high to be providing air support to an infantry unit, and none of the bombs or napalm canisters came close to the hilltop from which the fire directed toward us was coming. The Air Force jets then departed, leaving me in no better position than before they arrived. The enemy position on the hilltop was undisturbed. We were pinned down in the rocky riverbed and unable to move forward without exposing ourselves to enemy fire.

I was furious at the Air Force pilots. How could they be so incompetent as to miss the top of the hill with every single bomb they dropped? The Air Force is said to have a rule that they cannot fly lower than 2,500 feet because that is the altitude that ground fire can reach. Those are great rules of engagement if you are more concerned about your safety than accomplishing your mission. You stay outside the range of the enemy and use weapons that put him in your range of fire. I figured if the Air Force pilots did not have the balls to engage the enemy, they should turn over those F4's to someone who did.

The Marine Corps operates under different rules. The minimum distance at which to engage the enemy is face to face. After the failure of the Air Force to take out the enemy force pinning us down, the word was passed that we would have to handle this situation in the old-fashioned Marine Corps way. We were instructed to stage our packs and get light, fix our bayonets, and prepare to assault up the hill. The enemy was dug in, concealed, and had the advantage of favorable terrain. We would be running up a steep hill toward an enemy that had all the advantage. All of us knew that the casualty rate would be very high.

It might seem strange that the bayonet would be a viable weapon in the 20th century. However, in situations where you come face to face with an enemy in close quarters, using a rifle, or even a pistol, is dangerous. If you miss your enemy, the bullet may hit the Marine a few feet away but not in your sight. Plus, you never have to reload a bayonet.

I removed my pack and carefully placed it alongside those of my fellow Marines, fixed my bayonet, and verbally abused the Air Force. We were gathering for the briefing on how to conduct the assault on the hill. I don't know how many of us were going to make the attack. There were about 20 of us in my platoon forming up. I wondered how many of us would make it to the top and survive the engagement there.

Then, something truly wonderful happened. We got a call on the radio from a Marine F4 Phantom. The pilot, who was alone, told us that he had heard us on the radio with the other aircraft and wanted to know if we needed any help. He was returning to the airbase at Da Nang and still had ordinance! Hell, yes, we need help. The bad guys are on that hilltop on the west side of the river at the bend.

The camouflage-painted jet flew directly toward the top of the hill so low that it seemed that you could reach up and touch the it. The NVA, who had not even bothered to fire at the Air Force jets, opened fire with AK47's. As the plane neared the top of the hill, a canister of napalm was released. The plane was so fast and so low that the canister could not complete one full end-over-end rotation before impacting directly on the NVA position. That ended the fire from that position. The Marine F4 made one pass, dropped one munition, and flew off having fully accomplished his mission. I sheathed my bayonet, picked up my pack, and swore that if ever found that Marine pilot, I would buy him all the drinks he could put down.

We sent a patrol to the hilltop for battle damage assessment on the NVA position. The patrol found a scene reminiscent of Pompeii. Burned bodies were found in the positions they must have been in when the napalm struck the hill. From one of the bodies was removed a diary, which had not been destroyed by the fire that killed its owner. The dead owner was an NVA officer who had, according to the journal, been fighting in Vietnam for about 20 years. He had fought the French before the Americans arrived. The life of an NVA soldier was tough. While we Americans served tours of 12 or 13 months, the North Vietnamese and the Viet Cong were there for decades. Even we Marines assigned to rifle companies sometimes slept on cots in tents and had hot food. The NVA could not count on reliable food supply and lived in the mountains and forests. If an American was wounded, he would typically be medically evacuated by helicopter to a facility equipped to provide high-quality medical care. The NVA would have to rely upon medical care provided in the field and under far different conditions.

During combat, you are focused on what is happening around you. If you talk to several people who were involved in the same fight, they may have different stories to the point that you wonder if they were in the same battle. Everyone sees it from his own perspective, focusing on the threats nearest him. I was focused on the enemy position on the hill overlooking the river, from which fire was directed down to the Marines in the river bed ahead of me. I was unaware - and remained unaware for 50 years - of the more intense combat taking place forward of my position and out of my view.

At the head of our column, the 3rd Platoon of Bravo Company was in the point position. The point platoon had arrived at what must have been a beautiful scene, a waterfall on a jungle river. The platoon was ordered to go a couple of hundred meters beyond the waterfall, set up a defensive perimeter and wait until the remainder of the group caught up with them.

The river dropped off here to rocks below, and a mist rose from the bottom. But, the riverbed forward of the waterfall was dry. How could that be? As the Marines rappelled down the rock face on the edge of the falling water, they saw beneath them that the water was swallowed by an opening in the rock and disappeared underground.

After reaching the bottom of the falls, the Marines formed

up and moved forward. The flank squad and the point platoon would contact and engage the enemy first. Ahead of them, a large enemy force was watching from concealed bunkers and spider holes. The smaller group of NVA soldiers on the hill had not been the main ambush force. The enemy troops on the hill were in a position to fire upon the Marine formation from the side or behind while the Marines were in a vulnerable position., backed up in the steep-sided rocky riverbed while those ahead were slowed by the waterfall.

A standard Marine squad would have 14 members, comprised of three fire teams of 4 men each, plus a squad leader and assistant. A Navy corpsman would be a member of the squad and most likely a two-man machine gun team. During this time period, Marine infantry units were under-strength as a result of the very high casualty rate. A full-strength platoon would have around 43 members. In 1968, which was the height of the Vietnam war and the year with the highest casualty rate, a rifle platoon was at about half that strength.

Within a few minutes of arriving at the bottom of the waterfall, almost everyone in the platoon would be dead or severely wounded. As the point platoon moved across the open ground away from the waterfall, they suddenly received a massive volume of small-arms fire and rocket-propelled grenades. Many of them were killed or wounded in the initial fire. The Marines who were able to do so attacked into the fire, firing their rifles and throwing grenades as they advanced toward the dug-in enemy position. Their initial assault seemed to be having an effect, and the volume of fire from the enemy position in front of them decreased.

This ambush had been well-planned and was well-executed. The initial fire came from enemy positions in front of the Marines. As they advanced toward the enemy in front of them, another hidden enemy force on their right opened up. Soon only a few Marines remained standing, with most of the platoon struck down by the intense fire.

One of the Marines still standing was a machine gunner whose assistant gunner had been killed. He was firing into the enemy positions but needed help keeping his machine gun fed with ammunition. Each Marine, in addition to his personal gear, carried a belt of machine gun ammunition. The machine gunner yelled, "I need an A gunner." Sgt. Tom Prater, who was in charge of the platoon, ran to him and helped him load belts of ammo into the gun. He also began removing belts of ammo from dead and wounded Marines to keep the machine gun fire going.

Another Marine who had survived the initial ambush was "Duke." Duke was never called or referred to by his given name. To the Marines in his platoon, he was just "Duke." Duke was a curly-headed kid from New York City and was fond of doo-wop music, like the hit song, Duke of Earl. Duke would entertain the rest of the platoon with his doo-wop singing in the banana room, a sandbagged bunker in the 3rd platoon area made to function as a shelter during rocket and mortar attacks.

The Marines could not call for help. Their radio did not work. The night before, they had to sleep in the river, because the banks were too steep to climb out, and their radio had been soaked in river water.

Just then Sgt. Prater heard what he describes as "the

sound of angels." Jet aircraft flew overhead, obviously there to provide close air support to the Marines. However, Prater worried that the jets would see them and mistake them for the enemy.

Marines on the ground communicate with aircraft with smoke signals. Green smoke generally marks a friendly position. White smoke is used to mark a target. Prater asked if anyone had a green smoke grenade. One of the Marines said that he did, and Prater told him to throw it. The Marine threw the grenade, but it was not a green smoke ... it was white phosphorus. They feared that they had just marked themselves as a target.

White phosphorus burns intensely and is almost impossible to extinguish. Prater and Duke tried to smother it with mud, exposing themselves to enemy fire as they did so. Someone threw a green smoke grenade to mark their position as friendly. A jet flew by low enough to be seen to wave at the Marines, indicating that he understood they were friendlies.

Now the Marines knew that they could soon expect hell to descend on the enemy forces. The problem was that the enemy was close, and there were wounded Marines in the open field in front of the enemy position. Prater made the decision to try to retrieve the injured Marines and drag them away from the area where the jets would soon be attacking.

Prater and Duke ran into the field, and each picked up a wounded Marine, running back to an area of cover under intense small-arms fire. While retrieving one injured Marine, Prater noticed that Corporal Lonnie Gay, one of the squad leaders, was alive although terribly wounded. Gay was face down. A bullet had torn him open and his intestines were on the ground beside him, but he was conscious. Gay said, "Prater, get me next." Sergeant Prater promised that he would.

When Prater reached safety with the wounded Marine, he was exhausted. After a short rest, he resolved to bring back Lonnie Gay although he did not know how he would accomplish that. Gay was a big man, weighing 235 pounds without his equipment. Without knowing exactly how he would pull it off, Prater dashed out into the open field toward Corporal Gay. Duke ran beside him, determined to bring another Marine to safety. Once again, the Marines were running into the intense enemy fire.

Just a few feet from Gay, Prater felt as if he was hit with a baseball bat. The blow knocked him down. His first thought was, "why did Duke hit me?" He then realized that he had been shot. He was face down, and he could see his left foot alongside his head. The bullet had entered his left hip area and took out a couple of inches of his femur.

Duke, seeing Prater go down, threw grenades and fired his M16 toward the enemy position. The grenades and rifle fire caused the enemy to put their heads down long enough for Duke to attend to Prater. Duke pulled Prater close to a raised earthen dike that would provide a few inches of cover from the enemy fire. He pulled Prater's leg back to where it belonged and injected Prater with morphine. He then hurriedly scurried back to a position of cover as the NVA resumed firing.

Prater was now in the same predicament from which he had saved one Marine and had attempted to save Gay. He was severely wounded and lying helpless in front of a fortified enemy position, with only the cover of a dike in an old rice field to protect him and the potential to be hit by his own air support that would soon be bombing the enemy position just a few meters away.

The first jet came in very low. The short dirt dike provided protection from shrapnel and flying debris as bombs struck the nearby enemy bunkers. The blast wave jarred his body, causing him to grit his teeth so hard that he felt one of them crack. Each jet made two or three runs, dropping bombs and napalm precisely on the enemy position and avoiding the Marines trapped in the field in front of it. The heat of the napalm was intense. Prater heard enemy soldiers scream in pain while being burned alive. Then the jets were gone. But the enemy wasn't. Incredibly, the NVA were still there. The NVA were expert at surviving the impact of overwhelming firepower. They had fortified their position well. Spider holes allowed a soldier to be entirely underground in a hole with a top that would close, protecting him from all but a direct hit. As soon as the jets left, the surviving NVA were back in place, angrier than ever.

As Prater lay in the hot sun, unable to move from his position without being shot and physically unable to do so anyway, he could only wait. Duke provided cover with his M16 from behind a dirt embankment nearby. The machine gunner and another Marine acting as an assistant gunner covered the open field from a second position. Without Duke and the machine gun, the NVA could merely walk into the field and shoot him and the rest of the wounded. It was a standoff. The third platoon had gone ahead of the rest of Bravo Company, but Prater expected the rest of Bravo

Company and expected the rest of Bravo would come along soon. He did not know that Bravo Company was pinned down in the riverbed above the waterfall.

Prater saw a helicopter land and take off near the top of the waterfall, then return to land again. He could hear the fire of the enemy and see the rounds hitting the helicopter which soon lifted off and flew away.

Lonnie Gay, the enormous black Marine from Bakersfield, California, was, at 24, older than most of the members of the platoon. Prater and Gay were good friends. Both of them had been raised by grandparents. Lonnie's grandmother sent him food she made for him, which he shared with the other Marines in his platoon. Her pound cake was exceptionally good. Gay was a corporal and the squad leader of the 3rd squad of the 3rd platoon. He was not moving and was pretending to be dead so as not to draw fire.

One of the Marines in Gay's squad regained consciousness and was confused. He had been shot in the arm, but his legs were fine. He called out "where am I?" Prater responded to him, asking if he could run. The Marine said that he could. Prater told him that if he ran to where Duke was, he could find cover and might make it. The wounded Marine stood up, but when he saw that his arm was almost detached and the bone was showing in several places, he stopped to look at it. He was shot in the back and fell.

Gay attempted to crawl to the Marine, causing his intestines to be pulled out of his body. Prater told him to stop, but it was too late. The movement brought the attention of the NVA, who shot Gay in the back. Prater could hear the air escape from Gay's lungs as blood spurted from his body in arterial bleeding. The streams of blood landed on Prater. Prater watched helplessly as Gay died. Prater thought he saw Gay's spirit leave his body ... but then wondered if the morphine had caused him to imagine that.

Prater was dehydrated as he baked under the noonday tropical sun. Ants had been attracted to the blood splattered on Prater from Gay's spurting bleeding. The ants bit Prater, but the pain of that was much less than the pain from his hip and leg wound. He reached for his canteens and found that both of them had bullet holes in them and were empty. He called out to Duke, "I need water." With the water rationed at three canteens per day per man, water was in short supply. Duke threw a canteen of the precious liquid toward Prater, doubting that it would reach that far but giving it his best effort. He also knew that for it to benefit Prater, it would have to be close enough for Prater to reach it without moving his legs. To the amazement of both men, the canteen hit the ground and slid several feet right into Prater's hand.

The NVA were systematically killing the wounded. If a wounded Marine moved, they shot him. Prater had thus far managed to not come to the attention of the enemy. He was behind the paddy dike and just barely out of their view. Prater knew that when he raised the canteen to drink, they would see the movement and try to shoot him, but his thirst was so great, he did not care. He lifted the canteen and drank from it. Sure enough, the enemy soldiers noticed his raised canteen and opened fire. Ignoring the fire, Prater continued to drink until the canteen was empty. The NVA missed.

As Prater baked in the hot sun with the morphine wearing off and his pain increasing, I was fixing my bayonet along with other Marines who were preparing to assault the enemy position on the hilltop that was keeping the rest of Bravo Company pinned down and unable to reach the decimated 3rd platoon.

The sound of another jet caught Prater's attention. He looked up to see a single jet fly over his head very low and watched as a napalm canister fell from its belly. "Why is he bombing over there?" Prater thought. "The fight is over here."

As Prater had watched the life leave Lonnie Gay, he could feel his own life slipping away. The most important things in the world were his wife Ruth and their newborn son Tommy. Prater and Ruth had met when they were 13 and had been in love ever since. Prater had an inner voice telling him that he could let go of life or hang on to see his wife and son again. Now that the morphine was wearing off, the pain of his wounds was almost unbearable. Death would bring an end to that pain.

Prater realized that his death, which would bring an end to his pain, was likely. He was still losing blood and might bleed out before help could reach him and the other wounded Marines in the rice field. He also knew that the NVA would search the bodies and remove weapons and equipment as well as documents, maps or other items of intelligence value. Prater devised his last act of defiance to the NVA, a posthumous one.

Reaching to his pistol belt, he felt around until he found the smooth shape of an M26 fragmentation grenade. The 16-ounce high-explosive device, less than four inches long, was known as the “lemon” grenade to distinguish it from the “pineapple” grenade used in World War II, Korea, and the early days of the Vietnam war. Inside the lemon-shaped steel case of the M26 is a core of Composition B explosive, around which is a coil of notched steel wire. When the fuse sets off the explosive charge, the coiled steel wire breaks at the notches and is propelled outward. The lethal blast range of an M26 frag grenade is only about 15 feet. From 15 feet out to about 49 feet, the M26 is more likely to cause painful wounds than to kill. Hand grenades on the battlefield are far less lethal than those shown in Hollywood movies.

Those wounds cause the victim, distracted by the pain and shock of the grenade blast, to be less capable of repelling a follow-on attack or continuing an assault. A wounded soldier will require medical attention and will be less likely to be able to keep up with his comrades as they maneuver or retreat. Americans who were wounded by fragmentation could be airlifted to medical facilities by helicopter. The North Vietnamese Army did not have helicopters; they had to carry their wounded with them, slowing them down and limiting their movements. If you kill an enemy soldier, that eliminates one soldier from the battle. If you seriously wound an enemy soldier, you eliminate three: the wounded soldier and the two others that will have to carry him from the battlefield.

Prater carefully removed the grenade from his belt and placed it under his body, with the weight of his body holding the arming lever against the grenade. The arming lever, referred to as the "spoon," was a safety feature. The long slender metal strip held the firing pin of the grenade back from the primer. When the firing pin struck the primer, the fuse was activated and could not be stopped. Activating the fuse took two steps: pulling the pin and releasing the "spoon."

Prater, weak from loss of blood and incapacitated by his wounds, struggled to pull the pin of the grenade. He found that it was not possible, John Wayne movies notwithstanding, to pull the pin with his teeth. When the grenades were issued to us, the pins were relatively straight and would pull easily. For that reason, we always bent the ends of the pin outward to make it harder to pull and reduce the chances of the ring attached to the pin from snagging on jungle vegetation and causing an unintended, and almost always fatal, detonation. The extra effort required to pull the pin was never an issue during the adrenaline rush of battle. In Prater's weakened condition, it took all his effort to finally work the pin out of the grenade and place it under his body.

As long as the weight of Prater's body held the arming lever against the grenade, the internal fuse would not ignite. However, if the pin was removed, the only thing preventing the fuse burn and subsequent explosion was the weight of Prater's body. If he was lifted up or rolled over, an explosion was only seconds away.

Nearby lay the body of Leroy Palmer Jr., the physical opposite of Lonnie Gay. Palmer was a 20-year-old from Chicago. The only similarity was that both men were black. Whereas Gay was huge, Palmer was very small. His diminutive size gave rise to the nickname "Mouse," by which all the Marines knew him. He must have just barely met the size requirements to be a Marine. Mouse always had a smile on his face and a friendly greeting. Everyone loved Mouse.

Mouse was the only Marine in Bravo company that I

knew well. We had both arrived in Vietnam in November 1967. I had been assigned to participate in Operation Worth a couple of months earlier. Instead, I was re-assigned to mess duty for a month when the Marine in my platoon who had been assigned to that duty came down with VD after a trip to Da Nang. Mouse and I were both lance corporals at the time and subject to being assigned to some of the less pleasant duties. The two of us were responsible for the warehouse where the supplies of the mess hall were stored. We unloaded the trucks that brought the supplies and organized them on shelves until they were needed. Both the mess hall and the warehouse were tents with wooden sides. It was easy duty, and Mouse and I had time to chat and get to know each other. We became good friends. Mouse was a kind and gentle soul. Less than 13 months after he joined the Marines, his teenage wife Vernia would receive a visit from a Marine in dress blues, accompanied by a Navy Chaplain, telling her that she was now a widow.

The bodies of 21 Marines, some still alive, lay in the open field. Those who were still alive were either unconscious from their wounds or pretending to be dead. Usually, in such a circumstance, the NVA would move across the field killing the wounded and removing weapons and equipment from the bodies.

The battlespace was like a big chess board with significant penalties for losing the game. The NVA had taken out a substantial portion of the Marine force by their ambush of the 3rd platoon. They could not finish that job because Duke and the two-man machine gun team were preventing them from moving into the field. The Marine force, including the remainder of Bravo company, could not move forward because they were pinned down in the river bed by an enemy force overlooking the river. The Marines who had not yet reached the point in the river bed as to be under the direct fire from above still had the ability to maneuver. It was that group of Marines, me among them, that was given the mission to take out the NVA position on the hillside. As we neatly placed our packs and other equipment in rows on the jungle floor, double-checked our ammunition, grenades and other equipment in preparation to assault the hill, the Marine F4 phantom showed up and altered the chessboard.

After the complete destruction of the enemy position on the high ground overlooking the river bed, Bravo company moved forward to aid what remained of the decimated point platoon.

The advancing main body of Marines discovered the NVA position from which the ambush had been launched and understood the tactical situation. Prater heard the sound of helicopters and saw two Cobra gunships arrive and attack the enemy position with rockets. Duke took this opportunity, while the NVA had their heads down, to run to Prater. Prater told the astonished Duke that before he could be moved, it would be necessary to deal with the hand grenade under his belly. Duke reached under Prater, found the grenade and held the spoon to the body of the grenade while removing it from under Prater, then threw the grenade toward the enemy position.

Duke spread out a poncho and told Prater that if he could roll his body over onto the poncho, Duke would move his legs for him. The NVA was still shooting, and bullets were impacting the dike. Another Marine arrived to help Duke as they ran, dragging Prater on the poncho until they got to the dirt embankment that had shielded Duke during the firefight. Having his wounded legs dragged along the ground, hitting bumps in the field along the way, caused great pain to Prater. Still, under heavy fire, Duke and the other Marine threw themselves and Prater over the embankment to relative safety. Being thrown onto the ground once again caused Prater severe pain.

Help had arrived. Another helicopter, a twin-rotor aircraft of the type often used for medevac operations where several casualties are expected, approached and received a massive volume of fire. Bullets struck the helicopter as it hovered but did not land. A person jumped from the aircraft from an unbelievable height before it flew away, still taking fire. A stretcher was also thrown from the helicopter before it fled the fire. The person who seemed to have jumped from the helicopter was a Navy corpsman. Marines do not have their own medical personnel or chaplains. The Navy supplies staff to fulfill both those roles. Navy corpsmen are assigned to each platoon to attend to wounded Marines. Marines accept them as brothers and different from other Navy personnel.

The corpsman who reached Prater and began to treat his wounds turned out to be the same person Prater had seen jump from the helicopter in a tremendous act of courage based upon the altitude of the aircraft. In fact, the corpsman admitted, he did not jump. He had been standing in the door ready to jump when the helicopter reached the ground and fell out when the helicopter made an abrupt maneuver to avoid incoming fire. The corpsman examined Prater's wound, gave him a second injection of morphine, and said: "you're going home."

Other Marines and corpsmen were locating and treating the wounded survivors of the ambush, although enemy fire continued. A medevac helicopter once again approached to extract the wounded and take them to the hospital in Da Nang, 20 miles away. Prater and other wounded Marines were quickly loaded while the helicopter sat on the ground, rotors turning, taking fire. As soon as the wounded were loaded, the helicopter took off. However, the enemy fire had damaged the aircraft to the point that it would not make it to Da Nang. It landed a couple of miles away in a cornfield just outside the wire at the US Army Special Forces camp at Thuong Duc. Later, another helicopter transported the wounded from Thuong Duc to the hospital facility in Da Nang.

The following day was Sunday. A Navy chaplain conducted religious services in a clearing in the jungle. Seven body bags were lined up along the edge of the clearing, one noticeably shorter than the others. I had been saddened to learn that my friend Mouse was dead. I was told that he was shot in his small chest with a .50 caliber round. I was impressed by the fact that a chaplain was with us in the field. He must have been with us the entire time, humping the bush with the Marines. Not only was that duty physically arduous, but it was also uncomfortable and hazardous. We were deep into enemy territory, surrounded at times by an enemy force far superior in number.

The other six body bags contained the remains of James E. Bates, James C. Butler Jr., Lonnie James Gay, William Royal

Lindsey, Kenneth Ronald Semon, and Brian Eugene Campbell. In addition to those killed in action, 14 men of the point platoon and several others from the main body had been medically evacuated to Da Nang. Many of those, including Prater, would be sent on to the United States for long periods of recovery.

Eighty-nine US service members were killed in Vietnam on May 25, 1968. Thirty-one of them were marines. Seven were members of the 3rd platoon of Bravo company, the point platoon.

After the church service, a helicopter came for the bodies, and we heard the command to "saddle up." We were moving out to continue our sweep of the area searching for NVA to destroy. As we had just learned, they were not hard to find. Often, you didn't even have to look; they came to you.

EPILOGUE

I had seen only a small portion of the fight. I did not know what had happened to the point platoon. I just assumed the Air Force jets that made bomb runs that did not come even close to the enemy occupying the high ground over the river were trying and failing to hit that target. In fact, they were hitting their intended target: the enemy position that had ambushed and all but wiped out the point platoon I was under that mistaken impression for 50 years. I was angry with the Air Force that entire time.

While attending a reunion of the 1st Battalion of the 7th Marine Regiment, I was approached by a silver-haired man with a handlebar mustache. His name tag read "Tom Prater." He told me that Duke had asked him to look me up and say hello. Tom told me about how Duke had saved his life on May 25, 1968. He explained what happened at the waterfall and how the Air Force jets attacked the NVA force that had engaged the point platoon. For the first time, I realized my mistake in thinking that the Air Force had missed widely while attacking the NVA on the hillside.

MY DAY OFF

It was the evening before Memorial Day, 1968. The 1st Battalion of the 7th Marine Regiment had been in the bush on Operation Mameluke Thrust for weeks. We were in the normal state for grunts, tired and filthy. Our only baths for the past few weeks had been the result of wading during river crossings. Imagine our surprise and delight when we were told that we were going to get a day off for Memorial Day! We were told that we would be taking our day off at a place called Thuong Duc, next to an Army special forces post near an old French airfield. Unfortunately, it was a few miles away. The plan had been for us to be airlifted, but the rainy and overcast weather was not suitable for helicopter flight. It was, however, suitable for grunts to walk through. We walked all night to get there.

When we arrived at the air strip after "humping" all night, a light rain was falling. As soon as we left the jungle behind and entered the open space of the airfield, incoming mortars greeted us. The incoming mortars did not seem to be a big deal. They were falling on the other side of the field, getting closer and closer toward us as the enemy mortarmen adjusted their fire, but never getting close enough to be a real concern. After a while, the mortars stopped, at least temporarily.

We wasted no time in breaking out the small folding shovels we carried, known to the Marine Corps as "entrenching tools," and making ourselves homes in the form of fighting holes, along the airstrip. There was a cornfield next to us, from which I took a few corn stalks to make my hole more luxurious and comfortable. The corn stalks made an excellent carpet for the mud bottom of my home and were relatively soft. I was in the lap of luxury.

Some of the guys put up ponchos over their holes to keep out the rain. I did not. Those who did use ponchos over their holes were to regret it a few minutes later when the mortar attack resumed. I guess that the break in the mortar earlier attack was a result of the enemy breaking down their mortar tubes to move them closer, to bring them within the range to reach us. When the mortar attack resumed, they did indeed have the range. Shrapnel holes began to appear on the ponchos, which considerably reduced their effectiveness as rain shelters but did show the guys how close the shrapnel came to hitting them.

A fixed-wing aircraft landed on the field. I had never seen one like it. It had a very high tail and a rear ramp which opened to unload something, which I could not identify. It was a Caribou. It did not stay long, probably because its crew was averse to mortar shells.

A Marine ran past our holes, throwing each of us a Dixie

cup. He explained as he ran that the Dixie cups were for ice cream that would be coming on the next airplane. Ice cream!! What a treat that would be. I held on to my Dixie cup and occasionally turned it upside down to drain out the rain that had fallen into it. We waited, and we waited, and ... we waited. I must have turned my Dixie up upside down a few dozen times. Several hours later, the inevitable word came that the aircraft with the ice cream would not be coming. Perhaps the first aircrew had informed the ice cream delivery aircraft crew of the day-long mortar attack.

Meanwhile, we spent the day playing battleship. You remember that game where you lay out your assets on a grid and take turns with your opponent, designating squares on a paper battlefield where your rounds will land on his grid. If your rounds land on the grid square designated as the location of your opponent, you score a kill. If his rounds land on the grid square designated as your position, you are dead. The problem with this game of battleship is that you did not get a turn, and you really did get killed if your opponent picks the right square.

Fortunately, my square was not found by the enemy mortar crew. They did find some squares, resulting in 13 Marines being wounded. There were also a whole bunch of shredded US-issue green ponchos.

Our day off ended in the late afternoon when the word was passed to "saddle up" and prepare to move out. We moved out single-file into the jungle in the gathering darkness. As a 19-year-old lance corporal, I had no need to know exactly where we were going or what we would do when we got there. I just followed the Marine in front of me.

We walked all night to get back to the place where we had been the day before, to resume our operation. Then, of course, we walked all day to make up for time lost as a result of our day off. At least I was not walking flank. Delta Company had been assigned to move along the high ground to cover the movement of the main body. Night movement through the jungle was always dangerous. It was especially dangerous in this area that the NVA considered its home turf. Not long before midnight, a small group of NVA approached them, probably assuming that the Delta Marines were an NVA unit. In the ensuing fight, three Marines were wounded by grenade fragments and two NVA were killed.

The following morning at 0600, those of us who had taken our holiday at Thuong Duc arrived at our destination, which was a remote jungle location identified only by grid coordinates. That was the initial point from which to commence an attack on a location where an NVA force was thought to be. From that initial point, we moved out again into the mountains north of Thuong Duc, where huge numbers of NVA were thought to be. The enemy lived in those mountains. He had bases there in the jungle, which was so thick that airborne cameras could not see into it. That is where the enemy soldiers lived, trained and operated from. It was their refuge, the place from which they came to attack US and allied forces and the place to which they returned afterward. We were hunting.

A recon team code named "Prime Cut" was operating in that area, quietly searching for enemy forces and bases. We had with us an Army Special Forces advisor from the Thuong Duc base, along with two indigenous personnel and interpreters. It was more than a hint to us that were getting into an area with a large NVA force when the two indigenous personnel refused to advance and deserted.

It was a relief when we were able to set up a night position and take turns getting some rest. That did not last long, however. During the night, we were told to get ready to move out immediately. Word had been received, perhaps from Prime Cut, that the NVA had learned we were coming and had assembled a very large force that was on the way to intercept us and was close by. We needed to find a place more defensible than our current location. We needed high ground.

As I moved along the jungle trail in the darkness, I reflected on my day off. If I had the option of another day off, I would pass on it.

I much prefer my most recent Memorial Day celebrations. Hot dogs on the grill and cold beer beat playing real-life battleship while periodically turning your rain-filled Dixie cup upside down any day. Plus, there's a good chance that when someone promises you ice cream, you get actual ice cream instead of mortars.

OVERRUN

The hill was steep, and the going was tough. Much of the mountain was covered by triple-canopy jungle, the floor of which was thick and slimy mud, covered by rotting leaves and vegetation that made it even slicker. The June jungle heat sucked the moisture out of my lean body. The ration of three canteens of water per day ensured almost constant dehydration and the genuine risk of heat stroke, or worse, heat exhaustion. The humidity was close to 100 percent. My helmet and heavy flak jacket prevented the free flow of air around my body.

My M16 rifle was slung over my shoulder so that I could use my hands to pull myself up the slope from tree to tree. Along with my 130-pound frame, I was burdened by my rifle, bayonet, over 300 rounds of ammunition, two fragmentation grenades, two canteens, a first aid kit, a bolo machete and a pack with personal gear. My waist was too small to get enough magazine pouches on my utility belt to carry all 21 magazines of ammunition, so I carried 14 magazines in pouches on my belt and another seven in a bandoleer across my chest, Pancho Villa style.

I had learned a lesson about not carrying enough ammo. During a firefight, I reached for a fresh magazine, and my hand fell on an empty pouch. I remembered that I had reloaded once or twice so I thought there would be an empty pouch or two, but I reached for the next pouch and found it empty as well. Then I realized that all my ammo pouches were empty! I looked down at the bottom of my fighting hole and saw that my boots were covered with empty magazines. I had used all 14 magazines without realizing it. Being in a gunfight without ammunition is a sickening feeling.

The effort to climb the hill was exhausting, even for young men in excellent physical shape. During a 10-minute rest break, I looked around and saw that the other Marines were just as exhausted as I was. We were only halfway to the top of the hill. Our group included two rifle companies and the command group. Over 200 Marines were struggling up this hill. I considered what would happen if any one of us quit and could not or would not continue this punishing climb.

That was the moment that it became clear to me why the drill instructors at Parris Island had been so merciless in their demands of us. Quitting was never an option … never. If a recruit stopped during a run or forced march, the drill instructors would be all over him immediately, shouting and screaming or, if that did not work, kicking and punching. They told us that it is better to sweat in training than to bleed on the battlefield. The boot camps in San Diego and Parris Island were designed to weed out the recruits who could not handle the mental stress or physical demands of the Marine Corps.

On this jungle hillside in the suffocating heat, we were a long way from Parris Island. We were also in a very poor tactical position. It was essential to get to the top of the hill before dark and set up a defensive perimeter. Being trapped on the side of a mountain, with the enemy able to take the high ground above, would be a terrible situation. Every one of us had to be on the top of the hill by dark. The failure of any one of us to be able or willing to do that would put all the rest at risk.

Most people don't know what they are capable of because they've never had to push themselves to the maximum. The Marine drill instructors pushed each recruit until he thought he could go no further, then pushed him through the mental barrier to the next level of performance. When you think you can go no further, you actually have another 50 percent left in you; you just don't know it. The Marine Corps made you reach down and get that final reserve of strength. If you would not do that, you left boot camp without ever having worn the Marine Corps emblem or being called "Marine".

The men on Hill 1062 had all earned that right. It was a source of pride to all of those around me and one reason why, when the short break was over, we all picked up our gear and moved forward up a steep mud-covered jungle hill.

There were other reasons as well. To give up on a tough task would not only be letting yourself down, it would be failing to fulfill your commitment to the Marines who depended upon you and would be negatively impacted by your failure. If that was not enough motivation to succeed, there was the example set by all the Marines who had served during the 192 years since the Continental Marine Corps was established at a tavern in Philadelphia. Our drill instructors had made it clear to us that we inherited a reputation created by those who had come before us. We heard stories about the valor of Marines in iconic places like Belleau Wood, Iwo Jima, and the Chosin Reservoir. We all knew how "Iron Mike," a machine gunner in Korea, held his intestines in with one hand and fired a machine gun with the other hand to hold off a wave of Chinese soldiers. If we were allowed to join the club known as the United States Marine Corps, we were expected to uphold the honor of the Corps at all costs and without fail.

Walking into a Cloud

It took all day to reach the top of Hill 1062. As we neared the summit, we walked into what looked like a fog bank. We had actually walked into a cloud. It was an eerie feeling. My squad was assigned a section of the defensive perimeter to protect. Our exhaustion caused most of us to dig our fighting holes much shallower than we usually would. Soon the cloud moved away, but the moisture did not. It began to rain.

I shared a fighting hole with two other Marines in my fire team. I was the fire team leader. After digging our shallow hole, we used a poncho as a cover. After setting up a watch schedule, those of us not on watch dropped immediately to sleep. The Marine assigned first watch had used two of our three rifles to hold the poncho over our hole. He had fixed bayonets to the rifles and stuck the bayonets into the mud, converting the rifle to posts, then tied the poncho to the sling swivels on the butts of the rifles.

It seemed that I had barely closed my eyes when I was awakened to the sound of exploding grenades and gunfire. The top of the hill was thinner in vegetation, and the moon was shining on the poncho protecting our fighting hole from the rain. This caused the poncho to be a glistening square shape in the moonlight, clearly identifying our position. Our first action was to take the poncho down. I said more than a few four-letter words when I realized that the poncho was tied to the sling swivel on my rifle butt.

The gunfire and explosions were taking place on the other side of the perimeter. I saw no target to fire upon. Being concerned with fire discipline, I ordered the other two Marines not to fire unless and until I did. It is frustrating to be in a gun battle and not see a target. It is tempting to shoot into the jungle in an attempt to kill the enemy that you cannot see. However, there are a couple of good reasons to resist that temptation. The first is that once you fire, the enemy knows where you are. Not only does the sound of your rifle give you away, at night the muzzle flash will pinpoint your exact location. The concealment of your fighting position is a valuable asset that you lose with the first round you fire at night. The first round should kill the enemy; otherwise, you lose that asset. The second reason to exercise fire discipline is that you have to carry your ammunition, and when you're out, you don't know when you will be able to get more.

The Marines in the other fighting holes in our area of the perimeter did not resist the temptation to fire into the darkness. That brought return fire. Although the enemy did not know the location of our position, the fire drawn by the Marines around us caused bullets to fly directly over our position. We could hear the sound of M16's firing outgoing rounds, while AK 47 rounds were incoming, and made ourselves as small a target as we could. Then, there were the hornets.

When we set up our position on the perimeter, we took no particular notice of the dead tree behind us. We had no idea that the tree was home to a nest of hornets. We also had no idea how angry hornets get when gunfire slams into their home. Swarming out to investigate, the hornets were buzzing furiously around our position, adding to the chaos.

Behind us, the enemy had broken through our defensive perimeter, wounding or killing the Marines manning the positions in that section. The air was filled with a confusing mix of outgoing and incoming rifle fire, exploding grenades, bullets hitting the dead tree behind us, swarming hornets, and the painful cries of men hit by bullets and shrapnel. We stared into the darkness in our field of fire but never saw a target. One of the Marines in my fire team later told me that he saw someone run past our position but did not fire, because I had instructed him not to. If I had seen that person, I would have shot him.

Near the top of the hill, Sergeant Gorman had just removed his boots for the first time in weeks. Exhausted, he looked forward to getting some rest. A Marine for eight years, he was the NCO assigned to S-2, the intelligence shop for the battalion. Next to his position was the reconnaissance liaison NCO. As part of the command group, both NCOs remained near the battalion commander and senior officers so they could quickly relay information and accept mission tasking. Sergeant Gorman had not been asleep long when an explosion woke him up and sent him flying through the air. Stunned by the blast, Gorman located his helmet and .45 automatic pistol. Just before he was shot, a Marine yelled, "They are inside the perimeter!" The sound of gunfire and explosions of satchel charges and grenades were all around in the dark night. Gorman made out two figures moving toward him in the darkness. Recognizing that they were NVA, Gorman fired his pistol, hitting one of the figures.

In a different section of the perimeter, an enemy soldier came up the trail that we had made as we arrived at the top of the hill. On each side of that trail was a Marine in a fighting hole. With good fire discipline, they waited until the enemy was a few feet away and unable to see them in the dark. When they both opened fire, they emptied a magazine each into his body and saw tracers going through him. The enemy soldier stopped, turned around, lowered his rifle, and stumbled away down the trail. Expecting him to collapse and die on the spot, the Marines were astounded at his ability to continue after having been shot at least a dozen times at close range.

Near the center of the perimeter, a Gunnery Sergeant saw an enemy soldier approach and shot him in the chest with a .45 pistol. The powerful round knocked the enemy down, but then he got up, shot the pistol out of the hand of the Gunny, and ran off! The reason the 1911 Colt pistol, caliber .45, was used by the US military was because of its stopping power. To take a .45 round in the chest and get up and run away was remarkable.

The reason for the seemingly super strength and ability to shrug off bullet impacts would become apparent later when we examined the bodies of the dead North Vietnamese soldiers. They were "sappers." Our senior NCOs explained that sappers are basically suicide troops. They have tourniquets tied around their arms and legs to stop the bleeding before they suffer the inevitable wounds that will cause it. They are high on opium to kill the pain from the injuries they know they will receive. They are armed with a rifle and typically one 30-round magazine, in addition to grenades or satchel charges of explosives. Their mission is to breach their enemy's position and kill as many as possible, then make it back to their side to hand off their rifle and die. Apparently, they get some heavenly reward for that action.

Jungle warfare at night is not precise. It is hectic, confusing and scary. It is challenging to take out a sapper before he completes his mission. It is necessary to do neurological damage to disable him physically, because he will not bleed to death and is not immobilized by pain. You have to shoot the sapper enough times to get a hit in his brain or spinal cord. If you break the spinal cord, everything below that break stops working ... immediately. Most of the sappers were able to get away after inflicting their damage. A few were not. Those that managed to get out of our perimeter after expending their ammunition and explosives must have been wounded severely and most likely, mortally.

At first light, we discovered the dead bodies of seven NVA soldiers in green utilities. The evidence on the bodies of those who did not escape to die elsewhere explained why you could shoot sappers 30 times at close range or shoot them in the chest with a .45, with no immediately apparent result. I did not examine them carefully, but a Marine who was searching a nearby body showed me a small round tin container that contained a dark substance that looked like tar. He explained that it was opium.

I cannot say whether the sappers intended to die during their mission and were, therefore, suicide troops. I do not know what was in the mind of the enemy soldiers who ran into our position amid a couple of hundred Marines in fighting holes. I cannot imagine that they would think they had a realistic chance of doing that and not being shot and killed.

To recover enemy bodies after a fight was very rare. The enemy always tried to remove their dead from the battlefield. The enemy force was estimated to be of reinforced platoon size, perhaps 30 total. It is unlikely that there were any enemy survivors. The wounded sappers who managed to get out of our perimeter would have most likely bled out in the nearby jungle. Their comrades would try to find them, recover their equipment, and bury them in shallow graves.

Firefights are usually short, a few minutes at most. In most firefights, you feel no emotion, not even fear. You are focused on your actions. You respond to the situation without thinking and do as you were trained to do. That's how you end up firing all of your ammunition without realizing it. The fear, if that is what it is, comes later, after the battle. That's especially true if you have suffered casualties. You realize that you could have been one of them.

This engagement was different. It was taking on the feeling of something much more than a mere firefight. It lasted longer. The enemy had overrun our position and close quarter combat was playing out around me. We were far into hostile territory. We had been located and attacked by a hostile force that greatly outnumbered us. This was grim. As the first shock of the attack wore off, and I began to think about it, with NVA troops running through our position, it seemed unlikely that we could survive. Once our perimeter had been breached, there was nothing to stop the vast numbers of NVA troops in the area from entering our position and decimating our smaller force.

Illumination flares appeared overhead in an effort to light up the area and expose the movement of the enemy. However, the triple canopy jungle foliage obscured most of the light from the descending flares. As the flares drifted down under parachutes, they swung from side to side, causing the shadows in the jungle to sway and move in an eerie dance of light and dark.

I accepted the fact that I would probably not live to see the sunrise. Within the next few hours, I went through a range of emotions. Unlike most of the combat engagements I had experienced, I had time to feel fear. Then I was pissed off that these people intended to kill me. Finally, I accepted the fact that I was likely to die, and I made a plan.

I knew that if I were mortally wounded, I would probably live for a couple of minutes after suffering the fatal wound. I resolved that during those two minutes, I would be the meanest American these guys had ever seen. I would charge directly into them, shooting as many as I could and throwing my grenades until I could go no longer. They might take my life, but it would cost many of their own.

. . .

Meeting "Duke"

As the chaos and confusion waned, we turned to restoring our lines and tending to our wounded. My squad was ordered to restore the perimeter positions that were breached and remove the dead and wounded to a central location where we could protect the injured.

We moved to the place where the initial breach of the lines occurred and manned those positions. All the Marines in the holes where the breach occurred were dead or wounded. Corpsmen were treating the wounded as we arrived. My job was to help get the wounded back to the center of the defensive position and protect them. Some of the Marines who had suffered gunshot and shrapnel wounds could still fire a weapon, although they were in pain and would need morphine.

One Marine had a bandage across his eyes. He had been in one of the holes where the enemy broke through. I did not know the exact nature of his injuries, but it was evident that he was unable to see and therefore needed protection. I explained that he was inside the perimeter and surrounded by Marines who would protect him. He was unable to speak, perhaps because of the bandage on his face. We waited together the rest of the night. The wounded Marine used his finger to trace the outline of words on his palm: H-O-M-E.

At dawn, we were better able to take stock of our situation, which remained bleak. We needed to medically evacuate our wounded. We were due for routine resupply of C rations and water. We also urgently required plasma, morphine, and ammunition for severely wounded Marines who needed evacuation, or at least that plasma and morphine. However, we were surrounded by a large force of North Vietnamese regular troops. Helicopters could not get to us.

I overheard the battalion commander, a Lieutenant Colonel who was speaking on the radio and giving a situation report. It was his opinion that the sapper attack had been a probe to determine how tough a target we would be. There was no question that the enemy force was far superior in number. We were on their home turf. If they judged that we were a weak enough target, they would attack. If they thought the cost to them would be too great, they would not. He expected another attack, this time with a larger force.

We were on the edge of the range of our artillery support, which was provided by the 11th Marines. Artillery batteries were moved forward from their usual fixed positions in order to provide better coverage for us. Golf Battery and Kilo Battery were moved forward to Hill 52, then even further forward to the Thoung Duc airfield. Even so, we were also outside the range of artillery support.

The mountainous terrain made it difficult for us to utilize artillery support in some cases. Often the target would be on a ridge facing away from the artillery battery. Even high-angle fire from the artillery batteries could not solve that issue.

Listening Post

The night after we were overrun, my fire team was assigned to set up a listening post outside the defensive perimeter. The concept of a listening post is straightforward: the men assigned to the listening post slip out of the position and move to a location from which to listen and observe. The role of the Marines at the listening post is not to engage an approaching enemy force, but rather to detect the presence of the enemy and report it to the command center.

It was a dangerous duty for a couple of reasons. You were outside your defensive perimeter, where the enemy could find and engage you more readily, and your own troops could mistake you for the enemy. It provided the unique opportunity to be shot by the enemy in front of you or by your own troops behind you. If you detected the enemy and could return and report it, you would have done what you were sent to do. If the enemy found you or stumbled onto you and you had to engage, you would be in a firefight with people in front of you. When the enemy in front of you shot at you, the fire would go into the friendly forces behind you, who would then return fire, leaving you in the middle of a crossfire.

The three of us moved out in darkness into the jungle about 100 yards where we took cover and began to listen. Almost immediately, we heard enemy movement in front of us. The movement did not seem to be in any particular direction that we could identify. After listening for some time and hoping that the enemy was not moving toward our position, we returned to our lines and reported our observations. Thankfully, crossing back through a perimeter manned by trigger-happy and twitchy Marines went without incident.

Efforts to bring in supplies and evacuate the wounded by helicopter were unsuccessful. The helicopters took heavy fire from the NVA that surrounded us. A helicopter tried to drop a pallet of water onto our position but could not get in close enough. The pallet hit the edge of the hill and bounced down the hillside. The Marine Corps has a practice that the other services don't have. They assign pilots to ground forces to coordinate air assets. I think a secondary reason is to have the pilots see what ground forces need and how much they rely upon Marine air to get the job done, as well as the consequences of failing to provide adequate support. Our air officer was screaming at the helicopter pilot who dropped the pallet of water. In defense of the pilot, he was taking heavy fire. He was also trying to deliver something that a couple of hundred Marines needed to survive. There was a collective groan as the pallet of water bounced down the hillside.

By the day after we were overrun, all of us had eaten our last c ration meal. A millipede, long and black, with a hard shell and a zillion legs, crawled around the edge of my recently deepened fighting hole. I watched as his legs caused small grains of sand to slide from the rim down into my hole. I wondered how much meat there would be if you removed the legs and shell. If it had returned the next day, I would have eaten it.

Evacuating Duke

Duke and 26 other Marines, both dead and wounded, awaited the arrival of helicopters to take them away from Hill 1062. Some would be going to the morgue in Da Nang, known as "graves registration." The wounded would be triaged and sent to the Naval Support Facility in Da Nang.

The most seriously wounded would go to the hospital ship waiting just off the coast in the South China Sea. From the hospital ship, some would go to hospitals in Japan, and others, the most critically wounded, would be sent back to the U.S. to naval hospitals near their homes. Some would remain in hospitals for several months and never be fully restored to physical health.

The pain in Duke's eyes was intense. The morphine, carried by infantrymen for just this type of occasion, helped but it was in short supply. He ran his fingers under the battle dressing on his face and felt what he described to me as a runny egg. The grenade shrapnel had penetrated both his eyes. The last thing that Duke would ever see was the flash of an exploding grenade.

Epilogue

About 40 years later, some of us Marines who had considered eating things like millipedes on Hill 1062 were gathered in a banquet hall at our bi-annual battalion reunion, after enjoying an excellent dinner that did not include insects. I had always wondered what happened to that Marine who was wounded on Hill 1062 and had his face bandaged. I hoped that he made it home and recovered from his wounds. I asked my colleagues if anyone knew his name or his fate. Nobody did, but one of my friends said he might be able to find out.

Sometime after the reunion, I was contacted by the Marine who said he would try to find out about the wounded Marine from Hill 1062. He told me, "The Marine you are looking for is Abraham Laufgas. He lives in Florida, and he is blind." I was provided a phone number.

I was disappointed to learn that he had not recovered his sight. I had a vision of a blind veteran living in low-income housing, leading a miserable life. I called the number. When Abe answered, I introduced myself by name and began by saying that he probably did not know who I was. He cheerfully replied, "sure I do, Tony, you are a telephone solicitor." I explained that I had met him on Hill 1062 in Vietnam.

Abe, who had the nickname "Duke" in Vietnam, had returned to Miami after being medically discharged. The state of Florida gives preference in the issuance of liquor licenses to disabled veterans. Duke obtained a liquor license, which generated income. He invested the income and bought stock, which skyrocketed in value. He had always enjoyed rock music and had a business promoting rock bands from the 50's and 60's. I was relieved to hear that Duke was doing well for himself.

Duke married and raised a family. Unfortunately, tragedy would once again visit him. Duke and his wife had a daughter who was 5 and a son who was 2 when their third child, another girl, arrived. Two months after the arrival of their third child, Duke's wife, still in her 30's, suddenly died.

The authorities suggested that Duke, a man who was totally blind, would not be able to raise his children and for their safety, the children should be removed from his custody. Duke was having none of that. He set about learning how to change diapers, cook meals, and do the other things necessary to raise his children. With the assistance of the 5-year-old, Duke was able to provide his family with a safe and comfortable home. Duke is proud of all his children. His oldest daughter had just passed the Florida Bar exam when I spoke with him.

The most remarkable thing about "Duke" after his service in Vietnam, was the extraordinary accomplishment of raising his children.

THE SHOT THAT I AM GLAD I MISSED

The figure stood motionless in the surreal green glow of the Starlight scope mounted on my M14 rifle. The scope's crosshairs were centered nicely on its chest. My finger was on the trigger. This would be an easy shot. The figure was only 200 meters away, standing in an open field looking toward our small compound.

"Charlie" operated mostly at night. Darkness provided cover to move from one place to another, conduct reconnaissance, plant explosive devices, or conduct combat operations. The local villagers knew not to be outside after dark. Anyone moving about at night was assumed to be Viet Cong and would be shot. It was a firm rule and there were no exceptions.

What a dumb ass. Any Viet Cong dumb enough to stand in the open observing a nearby Marine compound, even at night, would quickly be removed from the VC gene pool. Perhaps, I thought, this guy has no understanding of night vision devices. The Starlight scope was a first-generation device that turned the black of night into an eerie green world where objects and people appeared as black figures without much detail. The shape of a person could be identified, but not the color of their clothing or skin, gender or other details. It was as if you were seeing the persons shadow. Although primitive, the starlight scope was a huge benefit in night warfare.

I was in a tower with a clear field of view. My rifle rested on a sandbag to hold it stable. I could not have asked for an easier shot. The figure seemed oblivious to the mortal danger and remained as still as a silhouette target on a rifle range. I was in no hurry. I could take my time, and I did. I considered a head shot that would kill instantly and more mercifully than a traditional "center of mass" shot. In the end, I decided to aim for the center of the chest. Less room for error that way. No doubt this VC will be totally surprised when the 7.62mm copper jacketed round slams into his chest with the impact of a small sledge hammer.

I began my trigger squeeze. Slow and steady. It would be almost impossible to miss this shot. If the trigger squeeze is smooth enough, the recoil and report of the rifle comes as a surprise, although you can sense when it is getting close by the amount of trigger slack you have taken up. My trigger pull had gotten to the point where the rifle would fire at any second when I heard a "boom." It was not my rifle. It was an outgoing mortar round fired from the mortar pit next to the tower.

The mortar crew had saved the life of my target, which darted into nearby bushes at the sound of the mortar firing.

Of all the luck. I could not have wished for a better setup. The target escaped the impact of my bullet literally by a second. I resolved to keep an eye on that area. The figure seemed not to know that I had been watching and might return. I expected to be in the tower at the bridge over the Tuy Loan river for another few weeks.

The Tuy Loan Bridge

The Tuy Loan River was a liquid barrier flowing lazily between Da Nang and the combat bases to the south. The peaceful setting belied the danger that lurked there. The villages that lined the rivers banks were under VC control. It was from the banks of the Tuy Loan that the first rockets were fired into Da Nang.

A bridge crossed the Tuy Loan on the road leading south from Da Nang, allowing trucks, tanks, and other vehicles to move back and forth between Da Nang and the areas to the south, where the Viet Cong lived and operated. The bridge was a critical part of the route along which supplies and personnel flowed from Da Nang to the combat bases to the south.

The duty of guarding the Tuy Loan bridge fell to a Marine platoon on a rotating basis. The wisdom of providing for the defense of the Tuy Loan bridge would become apparent in following months, during the interrogation of Senior Captain Nguyen Thanh Long, executive officer of the 141st Regiment of the North Vietnamese Army. Captain Long disclosed the plans of the NVA to attack the headquarters of the First Marine Division on Hill 327 in Da Nang.

In that plan, the 25th Sapper Company would infiltrate across the Tuy Loan and into Da Nang, where they would stage for the assault on Hill 327. A component of the plan was to attack and seize the Tuy Loan bridge.

As the mobile gun section leader of the 106 recoilless rifle platoon, I was sent to the bridge to determine if the small tower there would support the heavy rifle. It would not. It was unstable and rickety. I was surprised when my report of the unsuitability of the tower was met with orders to stay there anyway. We would be sent a heavy machine gun instead of a 106.

The Browning M2 .50 caliber heavy machine gun lived up to its description. It weighed 83 pounds. It was designed shortly after World War I by John Browning and entered service in 1933. It packed a hard punch and had a range of 2,500 meters, although its effective range was less. The ancient weapon would prove very effective in desert warfare, where a long reach was valuable, and would be in service well into the next century.

A couple of nights after sighting the mysterious figure, I got the word to keep an eye on the area of the front gate of the compound. Movement had been detected there the night before.

The next night, as I scanned the area with my Starlight scope- equipped M14, I observed a figure approach the front gate. This was even easier than the figure standing in the open. It was very close. Once again, I lined up my crosshairs on the chest of the figure and started my trigger squeeze.

Riflemen will understand when I say that your best shot surprises you. You concentrate on sight alignment and trigger squeeze. You only know that you have fired when you feel the recoil and hear the report. I maintained sight alignment on the chest of the figure and squeezed. Suddenly, the figure turned and ran! There was no reason for that behavior. There had been no sound to startle the figure, unlike the incident two nights before when the mortar was fired.

Damn again! I cursed my luck at two weird situations in a row that should have resulted in kills but did not.

I decided to take the shot. The starlight scope could not keep up with the movement of the running figure and blurred. As the figure ran, I mentally computed where the person would be, placed my crosshairs there, and fired. I figured I had a 50/50 chance of hitting the running target. The first- generation night division scopes were remarkable for their time, but they had. one huge problem: when you fired, the muzzle flash scrambled them, and they were unusable for several seconds afterward. I could not tell whether I had hit my target or not.

A squad was sent out to search. The searchers reported back by radio: no body, no blood, no drag marks. Damn. I missed!

After the mysterious visitor's repeated appearances, a couple of Marines were assigned to be at the gate to see if the VC returned. Our best guess was that a Viet Cong had been attempting to place a booby trap at the entrance of our compound. The two guys at the gate would prevent that and would kill any VC that tried it.

The next night, I heard an urgent sounding voice on the radio, probably meant for me, "Don't shoot, it's a kid." A Marine near the front gate had seen the figure close up and was afraid that I had also seen it and was preparing to shoot it.

The figure standing at our gate was an 11-year-old "retarded" girl! I use the term "retarded" because that was the term we used to describe her then. Perhaps she was just traumatized. She had no parents. She had no home. She had no idea that a couple of nights before, she had escaped death by a second because of the complete happenstance of a mortar team firing a round. Apparently, a rifle round passing so close to her that it must have missed her by inches did not deter her from trying to make human contact.

We brought her into our compound and "adopted" her. Marines wrote home with the unusual request for clothing for an 11-year-old girl. She became the Princess of our little compound. She had two dozen "uncles" and all the C-rations that she could eat.

In less than a month, we were relieved at that bridge by another Marine unit. We left our princess with the new unit, who adopted her as we had. I often wonder what happened to her.

Nothing short of divine intervention can explain what happened when twice, I was a second away from administering a death shot to an 11-year-old girl.

I do not doubt that the figure I first took to be an utterly incompetent enemy soldier standing in an open field was, in fact, this 11-year-old girl.

I am a warrior. I have killed enemy soldiers. I have not lost as much as a wink of sleep over that. If the squad that searched the area after I fired at the figure at the gate had found the body of an 11-year-old girl, my life might be different. I do not know that I could have put that behind me as a tragedy of war. Although it would have been logically understandable, it would have been hard to live with that on my conscience. I am glad that I don't have to do so.

I am glad that I missed.

LUNCH WITH THE AIR FORCE

Occasionally, the Marines from Hill 41 would make the 15-mile trip into Da Nang for administrative reasons or to make ammunition resupply runs. While in Da Nang, they would enjoy the rear area treats such as hamburgers at the massive base exchange at the Da Nang airbase. Da Nang was viewed as a rear echelon secure area akin to a resort. Marines were entitled to two periods of R&R (rest and relaxation) during their 13-month tour. R&R was a vacation from combat. One of those R&R periods was a week in places like Sydney, Bangkok, or Hawaii. The other was "in country" at a safe area within Vietnam. One such "in country" R&R center, where a battle-weary Marine could spend three days lounging on the beach or surfing, was at China Beach in Da Nang.

The weapons exchange

Another benefit of a visit to Da Nang was the ability to exchange older worn weapons for brand new weapons. The equipment used by Marines had often seen a lot of hard use and needed to be replaced, although this was rarely possible through the perpetually cash-strapped Marine Corps. The Air Force base exchange, which was the military equivalent of a department store, unintentionally operated an equipment exchange program.

At the Base Exchange, shoppers would hang their weapons on pegs provided for that purpose in an area outside the entry doors when they went in to shop. When they returned to retrieve it, they would find a worn, weather-beaten weapon in its place. Some Marine on a truck headed back into the bush would be admiring his brand-new rifle, which had only been fired a couple of times on a rifle range and had never seen mud.

When a Marine was issued a weapon, the serial number was recorded. That became meaningless because the first time the Marine was medically evacuated, his rifle would go into a pile that would be slung under the helicopter evacuating him. He would be taken to a medical facility, and his rifle would end up on a heap by the landing zone, where another Marine who was released from the medical facility would pick one to take back into the field. I expect that Air Force or Army personnel had some explaining to do when the weapon they were issued turned into a worn-out weapon with a different serial number.

The Salad Bar

During such an excursion, four Marines from Hill 41 decided to treat themselves to lunch at the Air Force mess hall. It was just like being in a restaurant back in the states. There was an excellent salad bar. The Marines were wearing their jungle utilities, flak jackets, helmets and associated "battle gear." As they began to enjoy the most excellent Air Force food, the Air Force Sergeant in charge of the mess hall approached and gruffly asked, "Who are you and what are you doing here?"

"We are Marines, we are here to eat," replied one of them.

"You can't eat here," said the Sergeant with an air of authority.

"Why not?" asked a Marine, taken aback by the refusal to serve them.

"You're dirty and you smell. You can't eat here until you take a shower and wash your clothes" was the haughty response of the Sergeant.

The Marines gave the Air Force NCO a defiant look. One stepped up onto the salad bar and started walking around in it with his mud-caked jungle boots. The others joined him. Lettuce, cottage cheese, and sliced beets went flying as the Marines destroyed the salad bar while commenting upon the paternity and wisdom of the red-faced mess Sergeant. Apparently, some property in the mess hall came to be damaged during this event. The Air Force Police were called, and the Marines were arrested for insubordination to the senior enlisted man, as well as the destruction of his salad bar and other equipment.

The protocol in this kind of situation was to take offending service members to their commands and turn them over to their Commanding Officer, explaining the charges. The CO would then take appropriate disciplinary action. The two Air Force Policemen drove the four back to Hill 41 to turn them over to their Commanding Officer. The drive from Da Nang to Hill 41 was not a pleasant ride in the countryside. The narrow dirt road was often mined by the enemy, and American vehicles, even in convoys, were sometimes ambushed by waiting Vietcong.

Hill 41

The Marines directed the Air Force driver to the small combat base on Hill 41. The Marine at the gate, surprised to see an Air Force vehicle this far from Da Nang, allowed the vehicle into the compound with a quizzical look, while noticing the Marines sitting in the back of the vehicle. The Marine combat base on Hill 41 was as foreign in appearance to the air policemen as the back side of the moon would have been. There was almost no vegetation. Trees and shrubs would have provided cover for enemy sappers if the wire was breached and the position overrun. Therefore, eliminating the foliage with a chemical defoliant knows as Agent Orange seemed like a good idea at the time.

The chow hall was a tent with wooden sides, rough wooden tables. There was no salad bar. Marines stood in the open and urinated into tubes coming out of the ground at an angle. They lived in tents with low walls made of sandbags or bunkers made entirely of sandbags. Near every tent, there were holes dug into the earth. The Marines called them "fighting holes." Marines were taught early in their training by their drill instructors that "foxholes" were places where foxes hid during trouble. Marines don't hide ... they fight. Thus, these holes, large enough for at least three men, were places where the Marines could find protection during enemy mortar or rocket attack.

Marines looked different from the airmen in Da Nang. These men were all lean, with almost no body fat, and were bronzed by the sun. Most of them walked many miles each day, moving silently and carefully along jungle trails and through rice fields to their ambush positions, where they spent the night quietly waiting for the Viet Cong or North Vietnamese Army troops to try to slip into the area. When the enemy did walk into the ambush, the sudden eruption of small arms fire announced an ancestral reunion.

The Marines slept little. When they were on night patrol, they were walking and attentive. When in an ambush position at night, they could take turns to grab a short nap. At periodic intervals, the PRC 25 radio would come alive with the voice of a Marine radio operator in the command bunker checking on the patrols. The voice on the radio would work through all the patrols: "Three Delta Four, Pennywise Delta ... if all is secure, key your handset three times." The radio operator of patrol Three Delta Four would press the transmit button on the radio handset three times, never speaking while on patrol.

The Marines ate C rations from green cans opened by the simple "John Wayne" folding can opener that most of them carried attached to their dog tag chain. There were only 12 different C ration meals, none of them particularly appealing. Typically, the patrol would stop on the way to their night ambush site and establish a 360-degree defensive perimeter. There, they would eat their C rations while waiting for it to become dark enough to move into their ambush position nearby without being seen by local villagers, many of whom would be changing from their daytime rice farmer persona to their Viet Cong persona after nightfall.

The command bunker was built of sandbags. Inside, radios were being monitored to stay in touch with the day patrols in the field and to receive information and orders from battalion headquarters.

The red soil produced red dust in the summer and red mud in the winter. When the Air Force truck stopped at the top of the hill and the Marines climbed down from the bed, they were met by the company Gunnery Sergeant, the senior enlisted Marine on the hill. After the Marines explained the situation, the Gunny brought the Marines and Air Policemen to the bunker where the Company Commander was working and asked them to wait while he arranged for the C.O. to see them.

The Marine Captain who was the commanding officer of Delta Company of the 1st Battalion of the 7th Marine Regiment, 1st Marine Division, was at work in the command bunker finalizing the night patrol routes when the air police arrived with the errant Marines. The airmen reported to the Captain and explained how the Marines had misconducted themselves, while the Marines stood at attention. The senior enlisted man on the hill, a Gunnery Sargent, stood in the corner of the small sandbagged room, listening. After a few minutes, he slipped out of the bunker.

The CO listened to the air police and told them that he

would take charge of the issue, then dismissed them. When the air police left, one of the Marines began to explain, "Skipper, we..." The CO raised his hand, motioning for silence. He had noticed the exit of the Gunny. In just a few minutes, the air police returned, visibly upset. "Sir, our truck is missing, someone stole our truck!"

"It can't have gone far. I'm sure you will find it. Just look around," advised the CO. The puzzled airmen left the bunker and began to search the small combat base. They wandered around the hill, asking Marines if they had seen an Air Force truck. Not surprisingly, none of the Marines had seen the truck. Finally, the truck was spotted behind a bunker at the back of the hill.

The relieved airmen quickly hopped into the truck and were happy to head back toward Da Nang where there were salad bars, NCO clubs, swimming pools, and movie theaters. With thoughts of happy hour at the club in their Air Force minds, they headed toward the only vehicle gate, which was the only way off the hill. It was late afternoon. As they approached the entrance, the Marine on duty closed it.

"Hey, open the gate, we need to get back to Da Nang," said the Air Force driver. The Marine responded, "gates closed." "What do you mean the gate is closed?" replied the driver, worry and confusion beginning to show in his voice. The Marine explained that the gate closes every day at this time and remains closed till the following morning. "But we have to get back, we can't stay here all night, open the gate!" The Marine gate guard was unmovable. The gate was closed to vehicle traffic and would remain that way till first light tomorrow morning. No further discussion.

The airmen watched as the combat base shifted from a more casual daytime routine to a more serious nighttime mode. Most of the action took place at night. The enemy used the concealment of darkness to conduct its operations, including slipping into the area with rockets. The Marines used the darkness to lie in wait along the jungle trails for the enemy to transit the area to set up their attacks. Marines, trying to conceal their relish for this unusual opportunity for amusement, were leaving the hill in single file, 15 meters apart, headed out on their night patrols. They were on their way to set up ambushes for Viet Cong and North Vietnamese troops who often attempted to slip into the area and launch their deadly 122 mm Russian-made rockets at the Da Nang airfield, as well as the salad bar, NCO clubs, and swimming pools.

The Gunny asked the CO, "Skipper, what do you want me to do with these Air Force weenies?" The CO responded, "Fuck'em, they aren't guests, put them on perimeter watch." On a combat base like Hill 41, most of the Marines are outside the wire patrolling. Several layers of concertina wire with sharp barbs surrounded the hill. Inside the wire circle were bunkers made of sandbags from which Marines, and tonight two air policemen, stood watch.

The air policemen were having an exciting, if unwelcome, experience that they must still talk about today. Perhaps they are telling people that they served with a Marine rifle company in Vietnam.

The C ration dinner they were provided was unlike anything served in the Air Force mess hall in Da Nang. The next morning, the Air Force vehicle was first in line at the gate. The Marine on duty explained that the gate would not open until they had received the report from the mine sweep team that was sent out from Hill 41 each morning to check the road for mines. The engineers often found mines that were planted the night before. Suddenly, being first in the line of vehicles did not seem like such a good idea.

GOODBYE VIETNAM

There were times, like the time we were overrun on Hill 1062, when I was certain that I would not make it out of Vietnam alive. There were other times when I thought that I probably would make it through my tour. There was never a time that I was confident of my survival.

Most of the time I was too busy to think about things like that. It is dangerous to let your mind wander when you need to be watching where you step and keeping your senses tuned. Once, while on an operation in very hostile territory, I walked out of the jungle into a small clearing on a mountaintop. I could see lush green jungle-covered hills below that rolled down to the South China Sea. Gray U.S. Navy ships floated in cobalt blue water under a clear light-blue sky. It was a strikingly beautiful scene. After about five seconds, I realized that I needed to get out of that clearing. The sudden realization that there were people around who wanted to kill me snapped me back into reality. It is hard to appreciate beauty in the presence of mortal danger. I looked forward to being in an environment where I would be free to enjoy the good things in life again without worry.

The midnight watch was a time when my mind was less busy. Unless we were on 100 percent alert, those of us on duty on the 106mm recoilless rifle took turns sleeping in the bunker while one of us stood watch by the gun. We had to be alert to the possibility of enemy infiltrators creeping through the wire surrounding the perimeter of our compound. When in a tower, we scanned the countryside with a starlight scope looking for enemy movement.

Not even a meditating monk can keep his mind free of thought. It was the midnight watch where I would sometimes think about being on a little hill surrounded by barbed wire in the middle of a hostile country in a heavily Viet Cong area. I would look at the clear night sky, seeing the same moon and the same stars that others could see from places where there was no need to worry about people trying to kill them.

I would think about how long I had left in my tour. Thirteen months is a long time, so long that it is hard to get your mind around when it will end. As you get closer, it becomes easier to see the end. Ironically, as you get near the end, you look back on more of your friends who have been killed or wounded in action. Sometimes it is best not to have time to think about those things.

Most Marines, and probably most other service members serving in Vietnam, kept a "short-timers" calendar. Some Marines made the calendar on their helmet cover, marking off months by putting a line through them. Others used the picture of a pinup girl, sectioned off into 395 squares, one for each day of their 13-month tour of duty. Each day, another square on the calendar was colored in until the long-awaited day when all the squares were colored, and it was time to board the "Freedom Bird" for home. The "Freedom Bird" was a commercial airliner with comfortable seats like those civilians enjoyed.

In late November 1968, after 13 months in Vietnam, I boarded the truck that would drop me off in Da Nang where I would board the aircraft we referred to as the "Freedom Bird," that took Marines who had completed their tours out of the country. I had often seen the truck pull into the compound loaded with new arrivals. When trucks left to take Marines to the "Freedom Bird," most of the seats were empty. It seemed quite natural to board a truck that was covered in red dust and had signs of battle damage. It seemed unreal that my tour was actually over.

The truck moved along the road to Da Nang, through the small villages with the rice fields where water buffaloes led by children pulled plows to farm the land.

Across the bridge on the Tuy Loan River where I had shot at the little girl. I had been assigned to a Combined Action Platoon in a village nearby, on this river.

There were a lot of memories on this road. It was on this road where one of my best friends died when a truck exactly like this one, going to Da Nang on an ammunition resupply run, struck a mine. It was across this road that I had fired at the NVA attempting to reach Da Nang during TET.

I remembered a Marine who had boarded a helicopter to be flown out of the bush at the end of his tour. The aircraft was shot down. He survived with only a broken leg. I would not be safe until we reached the city of Da Nang.

The farmers seemed not to notice the big green truck that produced the cloud of fine red dust as it kept up its speed headed into Da Nang. The locals knew not to get in front of the truck. It would not stop for any reason, including a civilian in the road. I watched for the last time men wearing black silk clothing resembling pajamas and women with black pants and white shirts working in the fields. Our truck was just a part of their lives, a part they would probably be happier without.

Most of the farmers and all of the young ones had never known life without war or the presence of foreign troops. Vietnam, Cambodia, and Laos had been colonies of the French and were known as French Indochina. In September of 1940, after France surrendered to Germany, Japan took control of French Indochina. Keeping the French colonial administration in place, the Japanese controlled the French Colonies, just as the Nazis controlled France and other French colonies, through the Vichy French government.

Indochinese communists organized to oppose Japanese rule. Ho Chi Minh, a Vietnamese communist, returned from China to lead the Viet Minh independence movement. Ironically, western Intelligence services, including the American Office of Strategic Services (OSS) provided assistance to Ho in his opposition to Japanese rule.

At the Potsdam conference in July 1945, Indochina was divided into two zones at the sixteenth parallel. The communists consolidated power and put down rival factions to take control of North Vietnam.

After Japan surrendered to the allies in August 1945, the Viet Minh launched the "August Revolution." Ho Chi Minh declared independence for the new Democratic Republic of Vietnam and established a degree of control in the North but met resistance in the South. Ho negotiated with the French for the semi-autonomy of a "free state" within the French Union. The south of Vietnam was not keen to go with Ho and objected, causing the agreement between Ho and France to fail. A year later, war broke out between the Vietnamese communists in South Vietnam and France.

The war with France, known as the "French War," lasted until the French defeat at Dien Bien Phu in 1954. Soon, the communists in South Vietnam began a guerrilla war against the pro-western government. The hostilities gradually escalated into what would come to be known in Vietnam as the "American war," in which I was a participant.

The farmers working in the rice fields along the road we were traveling probably did not know the historical reasons that they had always been surrounded by war. Most of them just wanted to farm their small rice field and be left alone. The Viet Cong had no intention of leaving them alone. The VC took the rice they produced, killed any village chief who opposed them, and forced the young men in the village to perform labor. The VC political officers explained that it was essential to fight the Americans who, just like the French, wanted to occupy Vietnam. Running off the Americans, like they had the French, would bring about the unification of the entire country and everyone would be happy in the Democratic Republic of Vietnam, just like their counterparts in the Democratic People's Republic of Korea. I don't think the political officer made it clear to the villagers that the people in North Korea were reduced to eating tree bark during famines and had zero freedoms, while those in free South Korea lived well and enjoyed their liberty. Neither the Democratic People's Republic of Korea nor the Democratic Republic of Vietnam was actually a republic or a democracy. They were brutal totalitarian communist regimes that offered no freedom and few choices.

When they took control of the city of Hue for three weeks during TET, the communists had demonstrated how they would treat South Vietnamese civilians if they took control of the country. The North Vietnamese rounded up and executed thousands of South Vietnamese civilians, including women and children. The NVA had come with lists of those who had shown some indication of opposition to the communists. Almost 2,800 South Vietnamese civilians were found in mass graves. Many more were missing and never found. A captured enemy document indicated that during the occupation of Hue, the communist executed 1,892 administrative personnel, 38 policemen, and 790 "tyrants." The families working in the rice fields had no doubt heard those stories.

The local population in our operational area assisted the Viet Cong, either because they were afraid to refuse assistance or had bought into the idea that the Americans wanted to occupy their country. I had no desire to occupy anything. I had come to Vietnam from DeFuniak Springs, Florida, to fight against those who would force their will upon the Vietnamese. If the people of South Vietnam wanted to become communists, that was fine by me. They could make that choice by vote, not by intimidation and bullying by communist thugs.

I had done my best for the people of South Vietnam. Now I was leaving, but they were not. They would continue to live in an environment where the VC would visit their villages to force them to provide food and labor and take reprisals against those that were suspected of assisting the Americans. The VC would plant explosive devices to kill or wound Americans who patrolled the area. The villagers would be blamed. The VC would recruit young women from the villages to be trained as nurses to render aid to wounded VC. They would have them leave home at night to go to the training. The Americans would detect them moving to their meeting and kill them, thinking them to be VC fighters

Being a Marine in Vietnam was tough. So was being a civilian in Vietnam. At least as a Marine, I knew when the war would end for me. That would be tomorrow.

As we arrived at the outskirts of the city, driving through Dog Patch, I began to get my head around the fact that I was actually done with my combat tour and would be back in the states soon. The land of hamburgers and milkshakes, pretty girls, and freedom from the stress of combat. Tonight, I would be in Da Nang, sort of a half-step back to civilization. There were movie theaters, bars with mixed drinks, good quality food, and almost no chance of being shot.

The Last Night

Finally, the truck arrived at the headquarters of the First

Marine Division in Da Nang. Other Marines were entering to be processed for departure. After the paperwork was done, the other Marines and I were taken to the Air Force facility to arrange our transport. After I was manifested on a flight out the next day, I was directed to a large open tent where those of us in transit would spend the night. I found a free canvas cot in the tent and stowed my seabag. The mood among the Marines assembling in the transit area was cautiously celebratory. We were all elated to be leaving Vietnam the following day.

A small group of us decided to look around the base and find somewhere to celebrate our last night in country. We found an Air Force non-commissioned officers (NCO) club. One of us was carrying two fragmentation grenades on his belt. When it was pointed out to him that the club would probably not let us in with the grenades, he solved the problem by putting them in the bushes outside the club. Even without the grenades, they refused to allow us in. Apparently, Marines were not welcome in Air Force clubs. We asked around and were told of another NCO club. We located the place but were once again refused entry. Now, we were getting a bit pissed off.

Members of all branches of service, except the Marines, were able to buy liquor at the base exchange and in bars. Apparently, there had been an unfortunate incident involving intoxicated Marines, which resulted in the rule that Marines were not allowed hard liquor. We had a small Army team on Hill 41 that would loan me an Army uniform when I wanted to buy liquor. Back in our combat bases, and even in the field sometimes, we were authorized two beers a day. However, the Air Force seemed determined not to let Marines drink on their base.

We learned of one more club on base. It was the "Acey-Ducey" club or 1-2-3 club, which served the three lowest ranking Air Force members. All of us were Corporals, above the rank that the club was supposed to serve. We all removed our rank insignia from our collars and walked in. The two young airmen at the door looked at us as if we were a motorcycle gang that had come to a church social.

Our uniforms were not dirty, but they were raggedy. We wore camouflage jungle utility uniforms that had seen a lot of tough wear. One Marine had replaced his failed bootlaces with the wire that came wrapped around cases of c-rations.

The airmen looked at each other as if they were trying to decide which one of them would have to tell us that we could not come into their club. One began to explain why we couldn't enter. He seemed to know that it was a lost cause and that we were coming in whether he wanted us to or not. Perhaps he had heard about the Marines who had been refused service at the Air Force salad bar and the consequences of aggravating hungry or thirsty Marines. We walked into the club. The two airmen stepped out of our way.

We took a table and ordered drinks from the bar. At tables around us, soft-looking young men in civilian clothes stole glances toward us. After a few minutes, an airman at a nearby table asked us, "What squadron are you guys from?" We told him that we were not in a squadron, we were Marines. Everyone in the club looked at us as if they had never seen a Marine, which may have been the case. "Are you the guys who patrol way out there and fight the VC that try to get to Da Nang or shoot the rockets?" When we affirmed that we were indeed those guys, free drinks began to flow. They wanted to know what it was like to be a Marine and listened wide-eyed to our stories while they kept buying us drinks.

I was drinking screwdrivers. I had 13 of them, the most I have ever drunk in one night in my life. I suspect that the drinks were pretty watered-down, perhaps because the bartender did not want to risk getting Marines too drunk. Tonight was indeed a time for celebration. My last night in Vietnam.

We made our way back to the transit tent and fell into our racks. About 0300, a siren sounded, and Air Force personnel began running toward sandbagged shelters. "We are getting hit," shouted an airman as he ran by. The sound of rockets exploding followed shortly. However, the explosions must have been a mile or so away. All the Marines went back to sleep.

The next morning we were led to a waiting commercial aircraft. We climbed the boarding ladder and entered the cabin, where we took our places in comfortable upholstered seats. Members of all services were aboard, all smiling. The flight attendants were women who seemed to have been selected because they were the age of the mothers of many of the passengers. Many of the men on the aircraft had not seen a non-Asian or "round-eyed" woman in over a year.

As the "Freedom Bird" left the runway, cheers arose throughout the cabin. The flight attendants smiled, having no doubt experienced this scene many times before. I looked out the window at the countryside as we climbed away. The beauty of the green jungle-covered mountains to the southwest belied their deadliness. In those mountains were North Vietnamese Army troops, planning, and training for their next operation against the Americans. One of those mountains was Hill 1062. Near Hill 1062 was a jungle river with a waterfall.

It was a much smoother and more comfortable ride than we had experienced arriving from Okinawa with our sore buttocks bouncing on the canvas seats of the C130.

My next stop would be Okinawa, where I would pick up the sea bag I had left there 13 months ago. It seemed like a lifetime ago that I passed through Okinawa heading west toward Vietnam, wondering what it would be like there. The experiences of those 13 months would be with me forever and in many ways would have a positive impact. I was literally a teenager when I had last been in Okinawa as a fresh-faced 18-year-old headed into a war, not knowing what role I would play in it or what to expect. Two weeks before my departure from Vietnam, I had turned 20. The difference between the fresh new guy going west and the tanned battle-hardened veteran going east seemed more like a decade than just over a year.

After processing through Okinawa, where I enjoyed a great steak and a couple of cocktails for $5 and got reacquainted with civilian clothing, I would be on my way to my next duty assignment. I flew into Pensacola, Florida, where I was met at the airport by the parents of a good friend from boot camp. My friend's father had been a Marine who fought the Pacific campaigns in World War II. The WWII Marine had cautioned me that when I passed through Okinawa, not to fall into any of the holes that he had dug there in 1945. During my time in Vietnam, I would get packages from him containing Wild Turkey whiskey in plastic aftershave bottles. It tasted a bit like aftershave, but I did not mind. Mixed with C-ration cocoa, it tasted great.

I enjoyed Thanksgiving dinner with that very kind family and felt at home with them. The next day, I took a bus to DeFuniak Springs, where I spent my two weeks leave time before heading to my next assignment. My orders were to report to the Second Marine Division at Camp Lejeune, North Carolina.

AFTER THE CORPS

When I returned from Vietnam, I had only eight months remaining in my three-year enlistment. After taking leave and visiting my friends in DeFuniak Springs and Pensacola, I reported to the Second Marine Division, Fleet Marine Force, based in Camp Lejeune North Carolina.

The Marine Corps did not know exactly what to do with those of us returning from the war. There were more than enough of us to fill all the positions in the rifle companies of the infantry regiments. Some of us were assigned as trainers for the new Marines being produced as rapidly as possible by the Marine Corps Recruit Depot in Parris Island. Others were sent to a variety of other jobs.

It was a time of active personnel movement. Marines were coming back from the war to fill slots vacated by those whose tours of duty were over and who were getting out of the Corps and starting new lives as civilians.

. . .

Military Police Company

I was assigned to the Military Police Company, not bad duty at all, and a gateway to my future career in law enforcement. Sometimes, things come about that seem coincidental but play a big part in your life. This was one of them. Some personnel clerk picked my name from a list of those returning from Vietnam to fill an open slot in the military police.

I don't recall having any training at all for my role in the military police. I was issued a .45 model 1911 Colt sidearm, a pistol belt, and a red armband with the letters MP before being sent out with another MP in a truck with a laughable little blue light on the roof.

We were waved down by a couple of Marines in a car. They were attempting to take a Marine friend to the hospital. He had apparently taken some drug that caused him to be so violent that they were unable to control him, and they feared damage to the car they were using to transport him. We were able to get him under physical control and took him to the emergency room of the base hospital.

We had him restrained with handcuffs on an examination table when the Navy doctor entered the room. The doctor seemed annoyed that we had handcuffed the patient and treated him so roughly. After all, the patient should be treated with dignity and kindness. The doctor admonished us for our treatment of the patient and directed us to take off the handcuffs. My MP partner and I had no doubt what would happen if we did that and suppressed our smiles as we took off the cuffs and stepped back. As soon as his hands were free, the patient punched the doctor in the stomach … hard. The doctor immediately changed his mind about how the patient should be treated and directed us to handcuff him.

Early Out

The Marine Corps maintains a constant presence around the world. There are US Navy ships designed as floating Marine combat bases. The combined Navy and Marine forces can deliver a landing party of Marines, ready to go ashore and fight, on any shore in the world in very little time. These units are positioned in strategically important areas.

The First Marine Division, based at Camp Pendleton California, is responsible for the Pacific. The Second Marine Division, from Camp Lejeune, covers Europe and the Mediterranean. Since I had seen very little of the world outside Florida and Vietnam, I liked the idea of visiting the ports of the Mediterranean. The fleet would return to North Carolina and swap out the Marine detachment after their tour was over.

I wanted to go on the next "Med Cruise." However, I did not have enough time left in the Marine Corps to complete the cruise. In order to go on the cruise, I would need to extend my enlistment. I went to the company office and told the company clerk that I wanted to extend my enlistment for three months. Without looking up, he subtly moved his head from side to side and hummed, "unh-uh. unh-uh."

I did not know what the clerk knew but was clear that he was telling me not to do it. I said, "well, maybe I don't want to extend" and left.

Two days later the announcement was made that those of us who had less than three months left, could be released early. The Marine Corps had come to the sensible conclusion that it was spending a great deal of money housing and feeding Marines who had too little time left to be useful. Returnees from Vietnam were stacking up. The Second Marine Division was overflowing with personnel. The money necessary to maintain those "short timers" could be put to better use in training, equipping, and deploying new Marines who could be sent to war. Since my enlistment would be up in July, I was eligible to be released in April. I had not given much thought as to what to do when my time in the Corps was up. The sudden shortening of my enlistment caught me by surprise. I had to make some decisions fast.

When the time came for me to be discharged, I stayed in the barracks for another couple of days, because I didn't really have anyplace to go. I had worked out that I needed to advance my education. Perhaps I would return to North Florida and enroll in a Junior College. Where would I work? There was not much available in my home town.

Perhaps I should re-enlist in the Corps. That would have been the easiest, but I felt that it was time for me to move on to the next adventure. I just did not know what that would be. I had never lived in a big city. The population of DeFuniak Springs was about the size of the crew on an aircraft carrier. I took out a map and looked around to see where the nearest big city was. Washington, D.C., seemed a logical choice from a geographical standpoint.

I asked around. A lieutenant who had lived in D.C. told me that he liked it there and recommended it.

About three days after my discharge, I got off a Trailways bus in downtown Washington, D.C., at about 2 a.m., with my sea bag of civilian clothing and all my possessions, and a little less than $500 mustering out pay. I had no clue as to what I would do there.

LESSONS

Life teaches you lessons. Some lessons are more painful than others. These are some things that my experience in Vietnam taught me.

Don't Watch the Tracers

During firefights I focused on my mission. It was my job to fire the 106 recoilless rifle against enemy targets, whether there was incoming fire or not.

Whether I was firing against a rocket site where rockets were being launched against Da Nang or shooting in support of Marines in ground action, I needed to be effective in delivering my fire. To do that, I could not be distracted by the worry of being hit by the bullets that I could hear and tracers I could see around me. I put them out of my mind.

You cannot deflect bullets by worrying about them. You will either be hit or not. Worrying about it only makes you less effective. The best way to keep from being shot is not to focus on your fear, but to kill the guy shooting at you. Be careful and deliberate in your actions. Take time to select a target and to aim carefully. Concentrate on hitting your target.

The same concept applies in other situations where there is a chance of an adverse outcome. I am not suggesting that you be reckless in your behavior or take unnecessary risks. There are times however, when you could choose to worry about something and have that worry immobilize you or rob you of the ability to do your very best to accomplish a goal. When you are in danger, physical or otherwise, concentrate on how to handle that situation most effectively. Don't watch the tracers.

You Can Do It

The difficulties faced by Duke in dealing with his injuries might have overwhelmed someone with less mental and emotional strength. The loss of sight was a significant issue, but Duke did not let it hold him back in being successful in business and establishing a loving family. Even through the loss of his wife and challenge of raising children while blind, Duke continued to conduct himself admirably, never giving up on himself or his duty as a father.

Keep things in Perspective

Hill 1062 was one of the most intense experiences of my life and taught me a valuable lesson. After you have been in a situation where you did not think you would live to see the sunrise again, the smaller problems in life take on a different perspective. Issues that most people regard as important, and that weigh heavily on them, seem relatively insignificant. That is not to say that lesser problems should not be a matter of some concern. However, the realization that things could be a lot worse can help calm you and keep your immediate problems in focus. Don't sweat the small stuff.

Don't Beat Yourself Up

It would have been tragic if I had shot the girl at the Tuy Loan Bridge. It could have turned out differently. I was doing exactly what I was supposed to be doing. Shooting Viet Cong. The fact that it was a child wandering around outside our compound in the middle of the night was unforeseeable.

I know veterans who are haunted by the consequences of what they did. It is important to realize that it is your intention that counts. If you take an action that has an unforeseeable result that you did not intend, you have no moral blame for it. If you act with honor and integrity, you are blameless.

You will sometimes be wrong

I thought that I understood what happened at the waterfall. After all, I clearly observed the Air Force planes miss the enemy position that was firing on me. Although what I saw and described was totally accurate, it was only half the picture. The planes were hitting their intended target, which happened to be a different enemy position.

There is a saying that you should believe nothing that you hear and only half of what you see.

Even if you personally witness something, you may be mistaken about what you think you saw. Keep an open mind to the possibility that things may not be as they seem.

I was wrong. My criticism of the Air Force was unfair. I believe that fairness is important, so I now officially acknowledge my mistake and apologize to the Air Force.

Keep your mind open to the possibility that you may not have correctly perceived things. You may be wrong. If you are wrong, admit it to yourself and others and reassess your views in light of the new reality.

MARINE GLOSSARY

Marines have their own language. Here are some terms that Marines in Vietnam would use:

C Rations

Combat Rations. Also known as "C rats". Food that Marines eat in the field. C rats come in green cans opened with a P-38 folding can opener usually attached to the Marines dog tag chain by running the chain through the small hole in the P-38. Pound cake and peaches were prized, Ham and Lima beans (also known as Ham and Mother Fuckers) were hated by most.

C4

A plastic explosive that comes in rectangular blocks. You can pinch off the amount that you need. It will not explode unless a blasting cap is used but it burns with extreme heat if lighted with a match. Used by Marines to heat C rations and boil water for C ration coffee.

DI

Drill Instructor. The Marine who taught the Marine to be a Marine. Before the recruit gets off the bus that brought him or her to boot camp, there will be a DI screaming at them. The screaming does not stop for 3 months.

E Tool

Entenching tool. A small folding shovel that Marines carry attached to their packs. The e tool is used to dig a fighting hole.

XO

The Executive Officer. Second in command of a Marine Company. Usually a 1st Lieutenant.

Freedom Bird

The airplane that took you out of Vietnam at the end of your tour.

. . .

Hooch

A Marine abode or living space. It might be a tent or a bunker constructed of sand bags. It is where the Marine keeps his gear, sleeps when off duty, and regards as his home.

Fighting Hole

Marines dig a hole that they can stand in so that only their head and shoulders are above ground, reducing the exposure of their bodies to enemy fire. The Army called them Fox Holes. Marines say that foxes hide in holes. Marines don't hide, they are fighting from that position.

Medevac

Medical evacuation. Also used to denote a helicopter used to extract wounded or sick from the field.

Mule

A small flat motorized platform with four wheels and a drivers seat that can be used to transport equipment. It is used as a platform for mounting a 106mm recoilless rifle, making it mobile.

NCO

Non-Commissioned Officer. Those service members holding the rank of Corporal or above but not commissioned officers who hold the ranks 2nd Lieutenant through General Officer.

Pop Smoke

Marines identified their positions to other Marine units and aircraft by the use of smoke grenades that came in different colors. Pilots could tell the wind speed and direction by the smoke trail. An approaching helicopter might ask the Marines on the ground to "pop smoke". The pilot never designated a color of smoke, the Marines on the ground would choose the color. The aircraft crew would say what color they saw, the Marines would verify that color. This procedure guarded against the enemy, who also had smoke grenades, from luring the helicopter to their position with the same color smoke.

Skipper

The Commanding Officer of a Marine unit of Company size, usually a Captain.

The World

The United States of America, where there were hamburger joints, girls and cool cars.

TET

The Chinese New Year. The biggest holiday in Vietnam. The term is also used to refer to the biggest battle of the Vietnam War, the TET offensive of 1968 described in "Here They Come" and "Happy New Year."

ACKNOWLEDGMENTS

I am sincerely grateful to the following people without whom this story would not have been published:

My editor, Anna Palmer Darkes, who as a staff attorney during my time as a Circuit Judge did the legal research and editing of my court orders, judicial findings and opinions, making me look smarter than I am. I am grateful for her continuing assistance in that regard.

The Marines of the 1st Battalion 7th Marines Vietnam Association helped greatly with the research necessary to make this story accurate as possible. In particular: Terry Kirkland and Victor Vilionis.

Marines involved in the events described in this book helped me fill in the gaps of my knowledge of those events. Special thanks to Ken Dohanish and Matt Petit.

I very much appreciate the assistance of some very talented beta readers who reviewed each of the chapters looking for errors that my editor and I may have missed: Ezra

Graham, Terry Kirkland, Judge Dorothy Russell, Scott Tack-till, Dr. Elizabeth Nelson, Maria Hinds, Susan Bryan, Bob Hill, Wanita Scroggs, Lance Zellers, Rob Schwab, Susan Eber, Amy Elliot, Christine Groves, Rob Bensko, Stacy Preston, and Scott Darkes. The inspector general of beta readers, Mack Palmer, gave the manuscript a final scrubbing.

Joseph M "Bish" Bischetsrieder kindly provided the image for the cover from a photograph he took while on patrol in Vietnam. Jill Muller did her design and photoshop magic to enhance it.

Thanks also to Patricia Toups who helped recruit beta readers from Stetson College of Law in Gulport, Florida, an excellent law school and very veteran friendly.

ABOUT THE AUTHOR

Anthony H Johnson is a retired Circuit Judge, former Navy Intelligence officer, Special Agent of the US Secret Service, and United States Marine. He lives in a small beach town in Florida with his golden doodle "Bosun," flies World War II airplanes and cruises his sailboat, "Magic Dragon."

He writes true stories about his experience as an eighteen and nineteen year old Marine in Vietnam and his years in law enforcement, where he served as a police officer and detective in Washington, D.C., a deputy sheriff in Pinellas County, Florida, and a Special Agent of the US Secret Service.

www.anthonyhjohnson.com

Made in the USA
Columbia, SC
16 June 2020

11390256R00119